PROPAGATION OF WAVES IN SHEAR FLOWS

WORLD SCIENTIFIC SERIES ON NONLINEAR SCIENCE

Editor: Leon O. Chua
University of California, Berkeley

Series A. MONOGRAPHS AND TREATISES

Published Titles

Volume 15: One-Dimensional Cellular Automata
B. Voorhees

Volume 16: Turbulence, Strange Attractors and Chaos
D. Ruelle

Volume 17: The Analysis of Complex Nonlinear Mechanical Systems: A Computer Algebra Assisted Approach
M. Lesser

Volume 19: Continuum Mechanics via Problems and Exercises
Edited by M. E. Eglit and D. H. Hodges

Volume 20: Chaotic Dynamics
C. Mira, L. Gardini, A. Barugola and J.-C. Cathala

Volume 21: Hopf Bifurcation Analysis: A Frequency Domain Approach
G. Chen and J. L. Moiola

Volume 22: Chaos and Complexity in Nonlinear Electronic Circuits
M. J. Ogorzalek

Volume 23: Nonlinear Dynamics in Particle Accelerators
R. Dilão and R. Alves-Pires

Volume 25: Chaotic Dynamics in Hamiltonian Systems
H. Dankowicz

Volume 30: Quasi-Conservative Systems: Cycles, Resonances and Chaos
A. D. Morozov

Volume 31: CNN: A Paradigm for Complexity
L. O. Chua

Forthcoming Titles

Volume 4: Methods of Qualitative Theory in Nonlinear Dynamics (Part I)
L. Shilnikov, A. Shilnikov, D. Turaev and L. O. Chua

Volume 18: Wave Propagation in Hydrodynamic Flows
A. L. Fabrikant and Y. A. Stepanyants

Volume 27: Thermomechanics of Nonlinear Irreversible Behaviours
G. A. Maugin

Volume 32: From Order to Chaos II
L. P. Kadanoff

WORLD SCIENTIFIC SERIES ON NONLINEAR SCIENCE Series A Vol. 18
Series Editor: Leon O. Chua

PROPAGATION OF WAVES IN SHEAR FLOWS

A. L. Fabrikant
Yu. A. Stepanyants
Russian Academy of Sciences

World Scientific
Singapore • New Jersey • London • Hong Kong

Published by

World Scientific Publishing Co. Pte. Ltd.
P O Box 128, Farrer Road, Singapore 912805
USA office: Suite 1B, 1060 Main Street, River Edge, NJ 07661
UK office: 57 Shelton Street, Covent Garden, London WC2H 9HE

Library of Congress Cataloging-in-Publication Data
Fabrikant, A. L. (Anatolii L'vovich)
Propagation of waves in shear flows / A.L. Fabrikant, Yu. A. Stepanyants.
p. cm. -- (World scientific series on nonlinear science. Series A : v. 18)
Includes bibliographical references.
ISBN 9810220529
1. Hydrodynamics. 2. Oscillations. 3. Waves. 4. Shear flow. I. Stephaniants, IU. A. II. Title. III. Series: World Scientific series on nonlinear science. Series A, Monographs and treatises : v. 18.
QC151.F335 1998
532'.51--dc21
98-29453
CIP

British Library Cataloguing-in-Publication Data
A catalogue record for this book is available from the British Library.

This book is printed on acid-free paper.

Printed in Singapore by Uto-Print

To our parents

Preface

After at least a century of research in such a basic issue as wave–flow interactions a number of classical books have been published giving us an excellent description of the subject.

Why one more book? First of all, we attempted to present the state-of-the-art theory of wave propagation in shear flows, with a slightly greater emphasis laid on the areas of research in which the authors took active part. Besides, we were also inspired by non-traditional or not so traditional views which have not been described so far or are hard to dig out of a growing number of scientific publications. We can point out several aspects that, we believe, make our book different from the others.

Specific wave processes. Wave perturbations in hydrodynamic flows are traditionally studied in the theory of hydrodynamic instability. Relatively less attention was given to the effect of flows on propagation of the waves generated by external sources although such problems were considered in a series of surveys and monographs as applied to geophysical hydrodynamics, aerohydroacoustics, etc. This book draws a primary attention to wave propagation processes with an emphasis on amplification mechanisms that may virtually explain existence of high-amplitude waves in many cases.

A qualitative approach versus a formal one. The principal goal of this book is to introduce a unified approach in presenting analytical or numerical results, related to oscillations and waves in shear hydrodynamic flows. A physical rather than a formal mathematical point of view on hydrodynamic phenomena is used to consider different problems in the theory of waves in fluids. This physical approach is based on separating and highlighting the most significant effects that may show up in various wave processes observed in geophysical and laboratory systems.

Basic models. These effects can be considered within rather simple models that allow for both theoretical investigation and experimental testing and permit detailed analysis and classification of elementary processes of the "wave–flow" interaction in fluid dynamics. These processes among others form a general picture of wave dynamics in shear flows which enables a researcher to study the relations between various wave phenomena and generate new ideas, stimulated by results from a neighbouring field of science. The book is concerned with the development and investigation of a number of such basic models for waves of different types in shear and vortex flows.

Method of analogy. Much attention is given to the plasma–hydrodynamic analogy which is an effective heuristic method of qualitative investigation into general mechanisms of wave amplification and absorption in flows. It permits natural classification of hydrodynamic phenomena and prompts further generalizations. Of paramount importance is also the concept of negative energy waves in hydrodynamics. This concept augments significantly the plasma–hydrodynamic analogy and allows for interpretation of many results in hydrodynamic theory of stability using a simple physical language. Besides, it gives a useful hint for the search of new instabilities.

We hope that the book may be a useful tool in the development of new trends in further investigations. It may be interesting to specialists in fluid mechanics, wave theory, physics of atmosphere and ocean, and other related branches of science. Besides, we believe that it may be useful to graduate and post-graduate students as a supplement to standard lecture courses.

Chap. 1 (except Sect. 1.8 and Sect. 1.9), Chap. 2 and Chaps. 4-7, as well as Sect. 8.2 and Sect. 8.3 were written by A. L. Fabrikant; Chap. 3, Sect. 1.8 and Sect. 1.9 were written by Yu. A. Stepanyants, and Sect. 8.1 and Sect. 8.4 are a joint venture.

The book includes materials published earlier in the review papers (Rabinovich and Fabrikant, 1976; Andronov and Fabrikant, 1979; Ostrovsky et al., 1983; Fabrikant, 1987; Stepanyants and Fabrikant, 1989) which were broadened significantly and supplemented with results of numerous related investigations.

The authors are grateful to all their co-authors of previous works which laid the foundations of this book. For many years we enjoyed a unique research environment, created in the Institute of Applied Physics, Russian Academy of Science. This stimulating environment was being supported by a wonderful team of enthusiastic researchers, whose scientific interests covered practically all sides and applications of the modern wave theory.

One of the authors (Yu. A. S.) is also grateful to the Université de Toulon et du Var (France) and to Prof. Ph. Fraunie personally for their hospitality and possibility to complete the work on this book during his stay at that University in 1997 within the framework of PAST invited professor position. We are also grateful to Nadezhda Krivatkina for the English translation of the manuscript and to Natalia Starodubrovskaya for her technical help in manuscript preparation.

The book was written with partial financial support from the grants INTAS (No. 94-4057), RFBR (No. 96-01-00585, 96-05-64476) and from the Ministry of General and Professional Education of the Russian Federation.

Nizhny Novgorod, December 1992 – October 1997
A. L. Fabrikant, Yu. A. Stepanyants

CONTENTS

Introduction

Interaction of waves and hydrodynamic flows is a major underlying process for a wide variety of phenomena that we observe in nature as well as for various technological applications. The theory of oscillations and waves in regard to hydrodynamic flows have several different aspects. Of paramount importance is the generation of wave perturbations. Nonstationary flows were considered as a primary generation source for waves of different types. We can mention here Lighthill's explanation of sound generation by turbulence (Zarembo and Krasil'nikov, 1966) and its analog for internal gravity waves (Gavrilov, 1987), Phillips' mechanism of surface wave generation by wind (Phillips, 1980), sound generation in some whistlers (Chanaud, 1970), etc.

Numerous works were concerned with the analysis of the self-excitation of waves in flows, which is one of the principal problems in the theory of hydrodynamic instability (Betchov and Criminale, 1971; Drazin and Reid, 1981). Mention should also be made of a broad class of problems of propagation of the waves generated by external sources and their interaction with hydrodynamic flows. Although unstable eigenmodes are only part of the wave perturbation spectrum, problems and methods of investigation that were developed in the theory of hydrodynamic instability may be employed in a general theory of waves in hydrodynamic flows. Note also that analyses of wave refraction and scattering in inhomogeneous flows and their use, in particular for flow diagnostics, are typical in the theory of wave propagation in the atmosphere, in the ocean, and in other real media (Gossard and Hooke, 1975).

Simplest hydrodynamic instabilities were investigated back in the nineteenth century for flows with tangential velocity discontinuity (a vortex sheet) (Rayleigh, 1945). Studies of perturbations in shear flows with a continuous velocity profile that were initiated by Rayleigh, Orr, Sommerfeld, and others turned out to be a very complicated problem. Its solution necessitated the development of an asymptotic theory of equations with singular perturbations (Lin, 1955), as well as of numerical methods allowing for detailed analysis of a broad class of flows (Betchov and Criminale, 1971; Gol'dshtick and Shtern, 1977).

The mechanisms responsible for wave amplification and absorption as a result of interaction with flows play a significant role in wave propagation in hydrodynamic flows. This problem has received due attention in the literature but most papers usually lack qualitative and simple physical interpretation of related effects. A physical approach to the problems of fluid dynamics implies picking out, thorough analysis, and classification of simplest models corresponding to elementary acts of the "wave–flow" interactions, that could be used for qualitative interpretation of the processes observed. In this book we attempt to fill in this blank. We develop and investigate simplest physical models for waves of different types in shear and vortex flows taking into account vertical stratificaton, horizontal inhomogeneity, rotation of the medium, and other factors encountered in real conditions.

One can distinguish two primary mechanisms of amplification (absorption) of the waves propagating in flows of ideal fluid. The first mechanism is, essentially, the so-called "over-reflection" (the term was introduced by D. J. Acheson (1976)) which consists in the amplification of reflected waves due to their interaction with negative energy waves in a moving medium. The phenomenon of over-reflection was investigated for acoustic (Miles, 1957; Ribner, 1957) and internal gravity waves (McKenzie, 1972), surface waves on water (Basovich, 1978), and electromagnetic and magnetohydrodynamic waves (Feijer, 1963; Lupanov, 1975). It should be mentioned that waves may also be amplified due to the change of the dissipation sign in a moving viscous fluid.

The notions of wave energy and momentum in a medium give rise to a number of questions that are related to wave-induced mean flow and its contribution to energy and momentum. In spite of a rather long history (McIntyre, 1981) these questions still have dark spots. Ambiguity in the contribution of energy and momentum to waves and background medium is encountered in different fields of physics of continuous media. In order to overcome these difficulties it is necessary to formulate the concept of wave energy and momentum that would meet a general physical representation of quasi-particles as elementary wave perturbations of eigenmodes of the medium and, at the same time, would take into account the nonlinear interaction of waves and induced mean flow, that is accompanied by the propagation of a wave train.

The other mechanism of the wave–flow interaction is resonant absorption or amplification in critical layers where the flow velocity coincides with the phase velocity of the wave. The effect of the wave–critical layer resonant interaction is well known in geophysical hydrodynamics, in particular, as one of the most significant mechanisms responsible for the absorption of internal gravity waves in the atmosphere (Gossard and Hooke, 1975). At the same time, only minor attention was paid to the investigation of wave amplification due to resonance particle energy in a critical layer. Investigations of resonant interactions should include analysis of the physical mechanisms of the processes in the critical layer and elucidation of the criterion determining the possibility of resonant amplification.

We would like to emphasize that calculation of such wave problems is a rather complicated and tedious task. Even solution of linear problems needs subtle analytical methods and computer facilities as a rule. Consequently, the efforts spared on overcoming mathematical difficulties in the development of a theory often determine a formal mathematical approach typical of many works on wave theory in hydrodynamic flows. Investigation into analytical properties of solutions of the corresponding boundary value problems or efficiency of numerical algorithms are frequently issues of primary importance. Such an approach is far from promoting the physical understanding of processes.

The analogy between the "wave–flow" interaction in hydrodynamic systems and the respective phenomena in electrodynamics, plasma physics, and electronics is a handy tool in intuitive understanding, qualitative comprehension of phenomena, and often in choosing optimal methods of calculation. This analogy was first mentioned by Case (1962) and later developed in different directions. It was found that some physical notions (negative energy waves, negative dissipation, the wave–flow resonant interaction, etc.) that lay the foundations of a theory of plasma instabilities are effective in hydrodynamic problems too. The plasma–hydrodynamic analogy was used to give a physical interpretation of results well-known in hydrodynamics (Lin's rule of integration path around resonant singular points in shear flows, instability of flows with inflection point in the velocity profile, Miles' mechanism of wind wave generation) as well as to investigate a number of new hydrodynamic effects: analogs of linear and nonlinear Landau damping in shear flows, quasilinear interaction and stimulated scattering of waves by particles for wind waves in the ocean, cyclotron absorption at sound scattering by vortices, and so on.

The analogy may also be reversed allowing for the investigation, for example, of some types of plasma instabilities employing known results from hydrodynamics. In this case, effects common for systems of different physical origins are also found thus enabling researchers to use a unified physical language which facilitates exchange of ideas and methods in different branches of physics.

Besides refraction and energy exchange between waves and linear shear flows, the processes of scattering by local velocity inhomogeneities may be highly significant for wave propagation in moving media. Flows with localized vortices are of particular interest in this respect. Such vortex flows are typical of various hydrodynamic phenomena (Batchelor, 1970). Vortices may radiate different waves and, besides, scatter absorbing or amplifying the incident waves. Wave scattering by vortex turbulence was broadly discussed in the literature (e. g., Tatarsky, 1967). However, the theory of wave propagation in vortex systems has many unclear points, due largely to insufficient knowledge of elementary processes of wave scattering by solitary vortices or by other vortex flows and of the origin of wave amplification and absorption at scattering by vortices. In this case, again, a useful tool is the analogy with electrodynamic and plasma problems which concern the interaction of electromagnetic waves with rotating objects and with curvilinear electron beams in electronic microwave generators, as well as cyclotron wave damping in magnetoactive plasma where electrons gyrate in a magnetic field.

High-efficiency scattering by vortices of acoustic, internal gravity, and other waves of hydrodynamic origin enables us to use this effect for remote sensing of flows. In particular, a method of acoustic probing at atmospheric turbulence has advanced significantly in recent years (Krasnenko, 1986). Remote acoustic sensing is of considerable interest for the investigation of localized vortices in both laboratory and real conditions. Methods of acoustic remote measurements in turbulent and other vortex flows have a number of advantages over conventional contact techniques or probing methods based on electromagnetic waves.

Note, finally, that in this book we do not expound the results of nonlinear wave theory in shear flows, which were presented in detail in the monographs (Drazin and Reid, 1981; Gollub and Swinny, 1984; Craik, 1985) and to which we refer our reader. We would like to emphasize, however, that the plasma–hydrodynamic analogy is also broadly employed in the investigation of nonlinear effects in critical layers (Andronov and Fabrikant, 1979; Reutov, 1982; Tsimring, 1982, 1983).

Chapter 1

Wave Energy and Momentum in a Moving Medium

The laws of conservation of energy and momentum are broadly used in investigations of wave motions in hydrodynamic flows. The definition of these notions, however, is concerned with certain difficulties that are often ignored. For example, calculations of momentum and energy, which are the quantities of the second order of smallness with respect to wave amplitude, must take into account the variations of average parameters of the medium, in particular, wave induced flows.

There arises a problem of unambiguous differentiation between mean and wave fields of physical variables. A similar problem was encountered in electrodynamics, that is, the celebrated Abraham – Minkowski controversy over the momentum of electromagnetic waves in the dielectric media (Ginzburg and Ugarov, 1976). The momentum of quasi-monochromatic waves has for a long time been a controversial issue in hydrodynamic problems (McIntyre, 1981).

Wave momentum is often an auxiliary concept in the formal solution of linear problems and its definition reduces to the choice of suitable terminology. However, an understanding of the physics of many linear wave processes in inhomogeneous media, as well as investigations of some nonlinear processes, demands a detailed analysis of the entire complex of hydrodynamic effects accompanying the propagation of waves in the medium. This refers, in the first place, to nonstationary induced flows that emerge in the emission, absorption, and scattering of waves. Besides, there is a need for a universal language (as in the general theory of oscillations and waves) that would define uniquely the energy and momentum of waves of different types independent of their origin. Numerous approaches to the definition of wave momentum employed today impede the exchange of ideas and results in different branches of physics.

This chapter is devoted to a detailed interpretation of the terms "wave energy" and "wave momentum" in the medium and to refinement of their physical meaning. Analysis based on comparison with classical hydrodynamic flows (Fabrikant, 1988) gives the definition of a quasi-monochromatic wave pulse that naturally meets the general physical concepts of quasi-particles as quanta of wave perturbations in the media and, at the same time, takes into account wave-induced mean flows which determine a number of nonlinear effects (Dysthe, 1979).

1.1 Wave momentum and quasi-particles

When investigating small perturbations, the dynamics of which may be described in a linear approximation, it is desirable, strictly speaking, to calculate energy and momentum within a basic nonlinear problem (Ezersky et al., 1981) and to take into account the changes in the mean values of physical fields that occur as a result of nonlinear generation of low-frequency perturbation. It is common practice, however, to use an average Whitham's variational principle (Whitham, 1974). An average Lagrangian of the original nonlinear problem may be expanded in odd powers of the wave amplitude:

$$L(a, \omega, k) = D_0(\omega, k)a^2 + D_1(\omega, k)a^4 + \dots \qquad (1.1)$$

Restricting the consideration to the first term, which corresponds to a linear approximation, one can determine the density of the adiabatic invariant (wave action) of the linear problem:

$$I = \omega \frac{\partial D_0}{\partial \omega} a^2. \tag{1.2}$$

This quantity has a sense of the number of quasi-particles within a quantum approach and is used in the determination of the density of quasi-energy, $\mathcal{E}$, and of quasi-momentum, $\mathbf{P}_w$:

$$\mathcal{E} = \omega I, \quad \mathbf{P}_w = \mathbf{k} I \tag{1.3}$$

The value of I does not depend on the choice of reference system, consequently it is convenient to calculate the quasi-energy $\mathcal{E}_0$ in a motionless medium where I is positive. Then, through Galilean transform, one can obtain in a medium moving uniformly at a velocity $\mathbf{U}$ the following expression:

$$\mathcal{E} = \mathcal{E}_0 + \mathbf{P}_w \mathbf{U} = \mathcal{E}_0 \frac{\omega}{\omega - \mathbf{kU}} \tag{1.4}$$

Note that, when $\omega - \mathbf{kU} < 0$, $\mathcal{E}$ is a negative quantity: it is in this sense that negative energy waves (NEW) are regarded in a moving medium (Ostrovsky *et al.*, 1986).

For a broad class of problems (linear problems, in particular), quasi-particles may be represented through the formulas (1.3), (1.4). The laws of conservation of quasi-energy and quasi-momentum follow from linearized equations of motion averaged over wave phase and are closely related to homogeneity and steadiness of the unperturbed medium (McIntyre, 1981). However, when investigating processes such as, for example, the wave–flow interaction, nonlinear effects of the self-forcing of waves, and the like we must refer to the original definition of the conservation of energy and momentum according to which the laws of motion do not depend on time and place. Besides, complete expressions for energy and momentum that follow from the governing (nonlinear) system of equations must be averaged over wave phase, and the wave part that is square with respect to amplitude is then extracted. The square terms that were omitted in linearization of the governing system may make a contribution that is comparable with the quantities $\mathcal{E}$ and $\mathbf{P}_w$. Fluid motions corresponding to these terms have a sense of wave induced flows.

A wave induced mean flow whose velocity is square with respect to amplitude will, evidently, make no contribution to kinetic energy if the unperturbed medium is motionless. As to the momentum of the wave motion, it is convenient to divide it into a quasi-momentum (some authors use the term *pseudo-momentum* (McIntyre, 1981) that is obtained in a linear approximation and a momentum of a wave induced flow. Typical features of these two components are well pronounced in gravity waves on the surface of a heavy fluid.

Two-dimensional gravity waves on deep water are described by a nonlinear boundary value problem:

$$\Delta\varphi = 0 \quad \text{for} \quad -\infty < y \leq \eta,$$

$$\left.\begin{aligned} \frac{\partial\eta}{\partial t} + \frac{\partial\varphi}{\partial x}\frac{\partial\eta}{\partial x} &= \frac{\partial\varphi}{\partial y} \\ \frac{\partial\varphi}{\partial t} + g\eta + \frac{1}{2}(\nabla\varphi)^2 &= 0 \end{aligned}\right\} \quad \text{for} \quad y = \eta, \tag{1.5}$$

where φ is the potential of the velocity $\mathbf{V} \equiv (u, v) = \nabla\varphi$, η is the vertical component of the particles' shift $\boldsymbol{\xi} = (\xi, \eta)$ from the equilibrium, $\nabla = (\partial/\partial x,\ \partial/\partial y)$, and $\Delta = \nabla^2$. Solutions of a linearized problem have the following form: $\varphi = \varphi_0 \exp(-i\omega t + ikx + |k|y)$ provided that $\omega^2 = g|k|$. In the absence of mean flow ($\langle\mathbf{V}\rangle = 0$, where the angular brackets denote averaging over wave period), the average Lagrangian velocity of particles [Lagrangian drift (Phillips, 1977)] along x can be written as

$$\begin{aligned} < u_L > &= < u > + < (\boldsymbol{\xi}\nabla)u > \\ = < \xi\frac{\partial u}{\partial x} + \eta\frac{\partial u}{\partial y} > &= < u^2 + v^2 > \frac{k}{\omega}. \end{aligned} \tag{1.6}$$

The average Lagrangian velocity decreases with depth as $\exp(2|k|y)$ and determines horizontal mass transport:

$$S_m = \rho \int_0^\infty < u_L > dy = \frac{\rho}{2k} < u_L(0) > . \tag{1.7}$$

This mass flux may also be obtained within the Euler description taking into account mass transport in the layer between the crests and troughs of the wave (Stokes transport):

$$S_m = \left\langle \rho \int_{-\infty}^{\eta} u dy \right\rangle = \rho \langle u(0)\eta \rangle \tag{1.8}$$

Stokes transport is definitively related to the quasi-momentum of the wave:

$$P_w = \int_{-\infty}^{0} < \rho u_L > dy \equiv S_m = \mathcal{E}\frac{k}{\omega}. \tag{1.9}$$

A mean flow induced by a quasi-monochromatic wave in an ideal incompressible liquid may be found in the second-order approximation with respect to amplitude by averaging the boundary value problem (1.5):

$$\Delta\Phi = 0 \quad \text{for} \quad -\infty < y < 0,$$

$$\left.\begin{aligned} \frac{\partial h}{\partial t} - \frac{\partial \Phi}{\partial y} &= -\frac{\partial}{\partial x} < \eta \frac{\partial \varphi}{\partial x} >, \\ \frac{\partial \Phi}{\partial t} + gh &= 0, \end{aligned}\right\} \quad \text{for} \quad y = 0, \tag{1.10}$$

where Φ and h are the potential and the surface deflection averaged over the wave period, respectively. Time derivatives may be neglected taking into consideration that the group velocity of waves is small compared to the phase velocity of the long-wave perturbations whose length is equal to that of the mean flow. We thus have a quasi-static problem of a flow under mass source distribution moving over the $y = 0$ surface, with the gradient of the mass flux, S_m, being the source of mass.

It should be emphasized that the induced mean flow is non-localized: the picture of stream-lines is determined by the entire structure of source distribution as well as by boundary conditions. A simple example is a chute of finite length in which a homogeneous progressive wave is radiated at one end and absorbed at the other (Fig. 1.1). The Stokes transport in such a chute is compensated by a horizontally homogeneous, induced counterflow, which gives a zero total mass flux in a stationary regime.

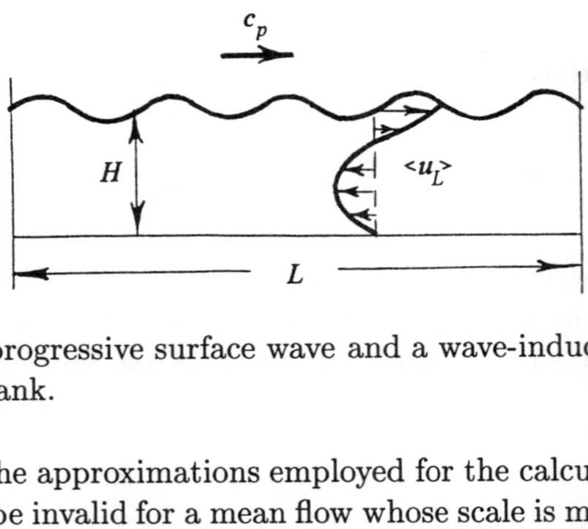

Figure 1.1: A progressive surface wave and a wave-induced counterflow in a bounded tank.

Note that the approximations employed for the calculations of gravity waves may be invalid for a mean flow whose scale is much larger than the wavelength. For instance, the velocity of the counterflow presented in Fig. 1.1 depends on the depth of fluid even if a deep water approximation holds for gravity waves (Ezersky and Papko, 1986). Besides, the characteristic Reynolds number for induced flow depends on the wave amplitude and if the latter is great enough the shear flow may become turbulent. Apparently, such a flow was realized in the experimental work (Russel and Onsorio, 1957).

The ambiguity of induced flow due to remote boundaries may be excluded by analysing dynamics of the wave trains having a finite spatial length l. If the depth of the fluid is $H \gg 1$, the stream-lines of a wave-train induced mean flow in an ideal fluid can readily be found by analogy with electrostatics: the stream-lines are exactly the same as in a two-dimensional picture of lines of force for the electrostatic field of unlike charges (Fig. 1.2) (McIntyre, 1981; Ezersky and Papko, 1986).

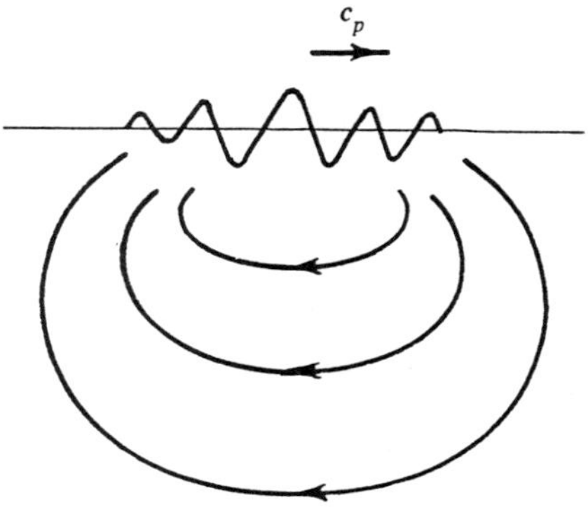

Figure 1.2: A mean flow induced by a train ofsurface waves on deep water.

A hydrodynamic object found in this fashion — a superposition of a surface wave train and the mean flow induced by this train— is a nonlinear generalization of the wave packet. The mean flow having the velocity proportional to the squared wave amplitude makes no significant contribution to the magnitude of energy but it must be taken into account in the calculation of momentum or mass flux.

It can readily be shown that the Stokes mass transport is exactly compensated by induced flow in each $x =$ const cross-section. Actually, from (1.10) in a quasi-static approximation we have

$$\frac{\partial}{\partial x}\left[\rho\int\limits_{-\infty}^{0}\frac{\partial\Phi}{\partial x}\,dy+\rho\left\langle\eta\frac{\partial\varphi}{\partial x}\right\rangle|_{y=0}\right]=0. \tag{1.11}$$

The first term in (1.11) stands for the mass flux through the $x =$ const cross-section in induced flow and the second term denotes the mass flux of Stokes transport.

Defining the total momentum of the medium involved with a progressive wave packet as an integral over the volume $\int \rho v \, d^3r$ we arrive at a conclusion that the total momentum of the gravity wave packet is equal to zero for the solution that is depicted in Fig. 1.1. The corresponding density of momentum, $\langle \rho v \rangle$, is not localized in the region occupied by the wave field; instead it is also distributed over the regions where the wave vanishes, in particular, for $k^{-1} \ll y \ll l$.

The difficulties enumerated above impelled McIntyre (1981) to renounce the notion of wave momentum and to use only the concept of pseudomomentum, which gives a complete and consistent description within a linear theory. We believe that the total momentum of a wave packet must be defined based on qualitative understanding of the physics of wave radiation and absorption in the medium and on comparison with the definition of momentum for classical localized flows in hydrodynamics.

1.2 Momentum of a localized incompressible flow

Many known flows of ideal incompressible fluid are localized flows: they are generated by the distribution of mass source vorticity in a bounded region. The flow is potential outside this region, with the velocity potential decreasing at large distances. This class of flows includes, in particular, potential stream-lining of bodies and localized vortices. One can readily determine mass transport in such flows. The total mass flux across an arbitrary given surface is, evidently, equal to zero for vortices because of their incompressibility. For a uniformly moving sphere, the mass flux across the surface σ is directed backwards, i. e., it is moving in the direction opposite to the velocity of the sphere, u (Fig. 1.3).

At the same time, calculation of the total momentum of flows is not an easy task. In particular, the integral

$$J = \int \rho v \, d^3r \tag{1.12}$$

taken over an infinite volume is not absolutely convergent (Landau and

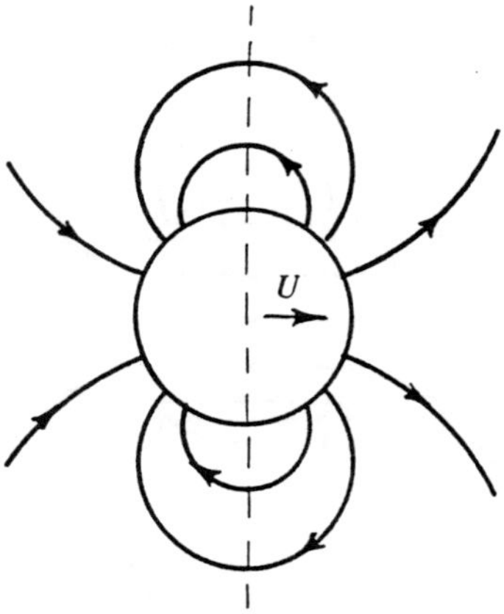

Figure 1.3: Potential flow around a sphere.

Lifshits, 1986) because the velocity potential at large distances has the form

$$\varphi = \left(\mathbf{A}, \nabla \frac{1}{r}\right) = -\frac{(\mathbf{A}, \mathbf{n})}{r^2} \tag{1.13}$$

and the velocity field drops as r^{-3}, which corresponds to nonintegrable singularity as $r \to \infty$.

Flow momentum may be found from the analysis of nonstationary processes involved with the variation of kinetic energy, $dE = \mathbf{U} \cdot d\mathbf{P}$. In particular, for a sphere of radius R we have (Landau and Lifshits, 1986)

$$\mathbf{A} = \frac{1}{2}\mathbf{U}R^3, \qquad \mathbf{P} = \frac{4}{3}\pi\rho\mathbf{A}. \tag{1.14}$$

Note that the integral (1.12) over the spherical layer of finite thickness around the sphere vanishes to zero. Consequently, the flow momentum (1.14) is specified by the contribution of an infinitely far region that is definitively determined by the dipole moment $\mathbf{A}$.

The physics of this result is understood if the potential flow around a sphere is replaced by an artificial flow generated by source and sink distribution over a sphere of radius R (Fig. 1.4). The flow has the form (1.13) in the outer region of the medium and can be represented

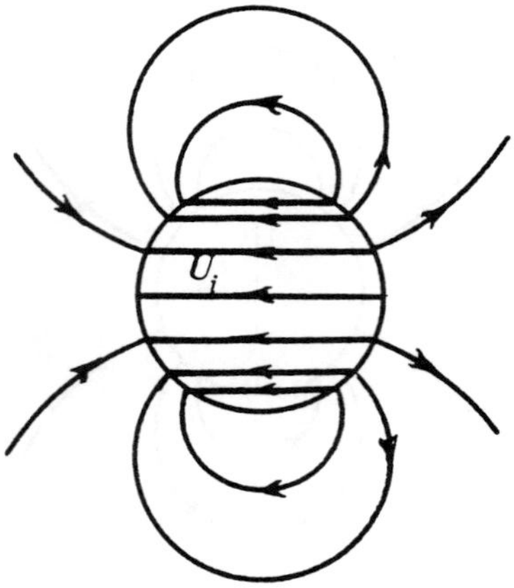

Figure 1.4: A potential flow emerging under the action of the sources and sinks distributed over the surface of a sphere.

as a uniform flow velocity, $\mathbf{U}_i = -\mathbf{A}/R^3$, inside the sphere. This field is a solution to a well-known electrostatic problem on polarization of a dielectric sphere in a uniform electric field (Jackson, 1965). Integration over the inside of the sphere yields

$$\mathbf{P}_i = \int \rho \mathbf{V}\, d^3r = \frac{4}{3}\pi\rho R^3 \mathbf{U}_i = -\frac{4}{3}\pi\rho\mathbf{A}.$$

At the same time, the contribution, $\mathbf{P}_e$, of the outer region to the total mmomentum is same as in the case of a stream-lined sphere, i. e., it is determined by the expression (1.14), and the resulting total momentum of the flow is

$$\mathbf{P} = \mathbf{P}_e + \mathbf{P}_i = 0.$$

Consider now an elementary pair of a point source and sink with a flow rate $\pm Q$ that are located at a distance l apart and form a dipole with a moment

$$\mathbf{A} = \frac{Ql\mathbf{x}_0}{4\pi}.$$

If the source and sink emerge simultaneously at time $t_0 = 0$, the flow is a superposition of the field of the point sink bounded by a shock wave

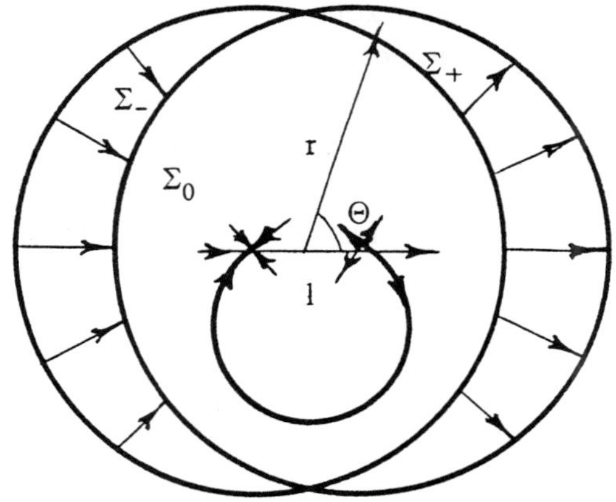

Figure 1.5: A dipole flow formed by the point source and sink appearing at the initial instant of time.

and the field of the point source bounded by an expansion shock wave (Fig. 1.5). When $t \gg l/c_s$ (where c_s is sound velocity), the flow inside the sphere of radius $r = c_s t$ coincides with the flow of ideal incompressible fluid and has a dipole asymptotic form (1.13). The layers, Σ_+ and Σ_-, formed by two eccentric spheres contain a velocity field

$$\mathbf{V}_e^{\pm} = \pm \frac{Q\mathbf{r}}{4\pi r^3}.$$

These layers determine the part of the dipole flow momentum which was at the "infinity" in an incompressible fluid approximation:

$$\mathbf{P}_e = 2\int\limits_{V_+} \rho \mathbf{V}_e^+ \, d^3r = 2\int\limits_0^{\pi/2} \rho \frac{Q\mathbf{x}_0 \cos\theta}{4\pi r^2} l \cos\theta \, 2\pi r \sin\theta \, r \, d\theta = \frac{4}{3}\pi\rho\mathbf{A}. \quad (1.15)$$

Arbitrary source distribution $Q(r)$ produces a more complicated flow, with the potential being determined by the Poisson equation

$$\Delta\varphi = Q(r).$$

But, given $\int Q \, d^3r = 0$, the potential (1.13) in the far zone, at a distance greater than the size of the source region, is determined by the total dipole moment $\mathbf{A}$. It is this far zone that contributes to the momentum $\mathbf{P}_e$ in expression (1.14) which determines the totalmomentum $\mathbf{P} = \mathbf{P}_e$ for a potential stream-lined sphere. For otherflows, the contribution, $\mathbf{P}_i$, of the source region must also betaken into account. In this fashion one can find, in particular, the momentum of a localized vortex having the vorticity $\zeta(r)$ for which we have (Batchelor, 1970):

$$\mathbf{A} = \frac{1}{8\pi}\int \mathbf{r}\times\boldsymbol{\zeta}\, d^3r, \qquad \mathbf{P} = \mathbf{P}_e + \mathbf{P}_i$$
$$\mathbf{P}_i = \frac{\rho}{3}\int \mathbf{r}\times\boldsymbol{\zeta}\, d^3r, \qquad \mathbf{P}_e = \frac{4}{3}\pi\rho A = \frac{\rho}{6}\int \mathbf{r}\times\boldsymbol{\zeta}\, d^3r \tag{1.16}$$

The foregoing remarks suggest that the momentum of a localized flow in an incompressible fluid must be determined by taking into account the reaction of an infinitely far region that is formed in the generation of the flow. Calculation of this reaction in an explicit form needs to take into consideration the finite velocity at which perturbations propagate. In particular, with compressibility taken into account, part of momentum is taken by the sound waves regenerated by the localized flow. However, the result depends neither on the type of propagating perturbations nor on the time evolution of the flow: the total momentum of the far zone is determined by the expression (1.14) and depends only on the dipole moment $\mathbf{A}$.

It should be noted that the need to take into account the contribution to the momentum of the wave perturbations propagating in nonstationary processes is not related to the divergence of the integral (1.12) in an incompressible flow in infinite space. For example, this integral converges: $\mathbf{J} = -4\pi\rho R^3\mathbf{U}/3$ if the sphere is stream-lined in a tube of finite cross-section (Benjamin, 1963). This quantity, however, does not describe the total momentum but rather the mass transport of the fluid that is, evidently, negative because the fluid is displaced by the sphere and takes its place, while acoustic perturbations, which arise as the sphere is accelerated and drift along the tube, contribute to the total momentum; the compression pulse goes forwards and the rarefaction pulse goes backwards (Fig. 1.6).

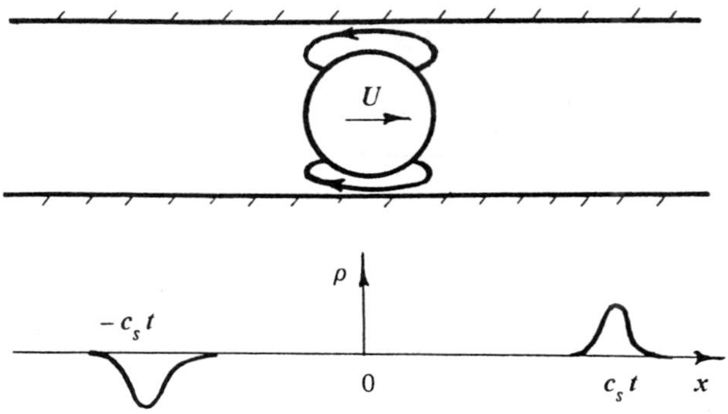

Figure 1.6: A flow around a sphere in a tube.

1.3 Momentum of a wave-packet induced flow. Long-wave radiation

The proposed approach is also valid for the localized flows induced by a quasi-monochromatic wave trains. Consider the simplest model of surface gravity waves where the length of the wave packet

$$l \gg H \gg \lambda = 2\pi/k$$

so that a shallow water approximation holds for mean flows while the propagation of gravity waves may be considered in a deep water approximation. Wave generation is accompanied by the emergence of source at the leading slope and of sink at the trailing slope of the wave packet. The distribution of sources is involved with Stokes mass transport and their intensity is determined by [see (1.9)]:

$$Q_m = -\frac{\partial S_m}{\partial x} = -\frac{\partial P_w}{\partial x}. \tag{1.17}$$

The formation of source and sink stimulates radiation of the surface level jumps (Fig. 1.7)

$$\Delta h = \frac{1}{2\rho c_H} \int\limits_{x-c_H t}^{x+c_H t} Q_m \, dx$$

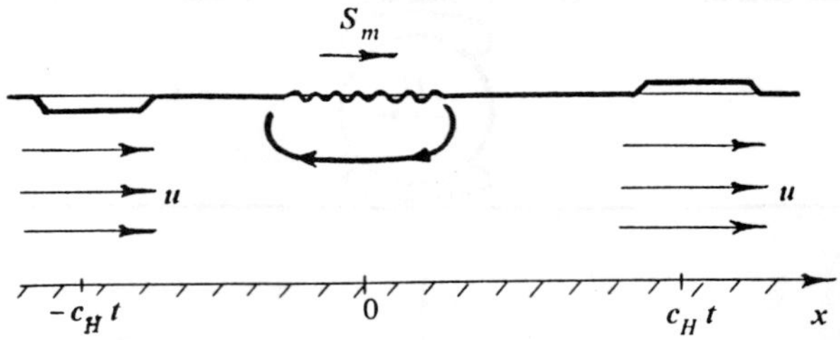

Figure 1.7: A wave train and an induced flow on shallow water.

which, in a shallow water approximation, propagate without dispersion at a speed $c_H = \sqrt{gH}$ (Landau and Lifshits, 1986). The flow velocity in an incompressible fluid under the surface elevation wave is given by

$$u = \frac{\Delta h c_H}{H} = \frac{1}{2\rho H}\int Q_m\,dx = \frac{p_w}{2\rho H}.$$

The perturbations receding away from the wave train with the scale of the envelope carry off the momentum

$$p_0 = \rho\int_{-\infty}^{+\infty} Hu\,dx = \int_{-\infty}^{+\infty} p_w\,dx. \tag{1.18}$$

Thus, although the Stokes transport is completely compensated by induced counterflow and the mass transport through an arbitrary cross-section is zero, the total momentum, with long-wave perturbations rapidly receding away taken into account, is equal exactly to the quasi-momentum (1.18). The momentum of force needed to excite a wave train must be precisely equal to the value given by (1.18). Then the momentum and energy of radiated quasi-monochromatic waves may be calculated within a linear theory neglecting induced flows.

An analogous picture occurs as $H \to \infty$. In this case, the flow depicted in Fig. 1.2 has a dipole moment

$$A = \int_{-\infty}^{+\infty} xQ_m\,dx.$$

The momentum of the flow consists of two portions. The first one is contributed by the region in the neighbourhood of the wave packet and is equal to zero because the Stokes transport is compensated by induced counterflow. The second portion is the contribution of the far zone which, like in the stream-lined sphere, is due to receding long-wave perturbations. Apparently, long gravity waves radiated at the instant the packet is born are such perturbations in this case.

Consider the radiation of long surface waves on deep water in the generation of a quasi-monochromatic wave packet. This process is described by system (1.10) in which the right-hand side of the equation for h may be expressed through the mass flux of Stokes transport S_m [see (1.18)]:

$$\frac{\partial h}{\partial t} - \frac{\partial \Phi}{\partial y} = -\frac{1}{\rho}\frac{\partial S_m}{\partial x} \quad \text{for } y = 0. \tag{1.19}$$

Let the wave packet be formed at the moment $t_0 = 0$ so that

$$S_m = S_m(x)H(t), \tag{1.20}$$

where $H(t)$ is the Heaviside unit function. Passing over to a Fourier transform

$$S_m(x) = \int S_{mk} e^{ikx} \frac{dk}{\sqrt{2\pi}},$$

$$h = \int h_k e^{ikx} \frac{dk}{\sqrt{2\pi}}, \qquad \Phi = \int \Phi_k e^{ikx} \frac{dk}{\sqrt{2\pi}}, \tag{1.21}$$

we obtain from (1.20) the equation

$$\frac{\partial^2 h_k}{\partial t^2} + \omega_k^2 h_k = -\frac{ikS_{mk}}{\rho}\delta(t), \tag{1.22}$$

where $\omega_k^2 = g|k|$. Solution to this equation may be readily found; for h_k

and Φ_k we have:

$$
\begin{aligned}
h_k &= \frac{ikS_{mk}}{\rho}\frac{\sin\omega_k t}{\omega_k}H(t), \\
\Phi_k &= -g\int_0^t h_k\,dt = -\frac{ikS_{mk}}{\rho|k|}\left[1-\cos(\omega_k t)\right]H(t).
\end{aligned} \tag{1.23}
$$

Coming back to space variables we express the potential of the mean velocity field in the form

$$
\Phi(x,y) = -H(t)\int_{-\infty}^{+\infty}\frac{ikS_{mk}}{\rho|k|}(1-\cos\omega_k t)e^{ikx+|k|y}\frac{dk}{\sqrt{2\pi}}. \tag{1.24}
$$

We are now in a position to write the total momentum of meanflow:

$$
\begin{aligned}
P = \rho\int_{-\infty}^{+\infty}dx\int_{-\infty}^{0}\frac{\partial\Phi}{\partial x}\,dy &= \rho\int_{-\infty}^{+\infty}\delta(k)\frac{ik\Phi_k}{|k|}\,dk = \\
&= S_m|_{k=0}\left[1-\cos(\omega|_{k=0}t)\right]H(t) = 0.
\end{aligned} \tag{1.25}
$$

Taking into account only the first component in (1.24) corresponds to a quasi-static approximation when the time derivatives in (1.19) are omitted. This part of the solution was considered in Sect. 1.1 by analogy with electrostatics (see Fig. 1.2). According to (1.9) the momentum, P_i, of this flow completely compensates the momentum associated with the Stokes transport, i. e., $P_i = -P_0$.

The second term in (1.24) does not contribute to the integral when $x, y \le l$ (the near zone), given $t \gg \sqrt{1/g}$, because here $\cos\omega_k t$ rapidly oscillates with the variation of the function $S_{mk}\exp(ikx)$. The second term in (1.24) describes the long waves receding in the far zone ($x, y \gg l$). According to (1.25), they make a contribution of $P_e = -P_i = P_0$ to the total momentum.

Note that the surface perturbation, $h(x,t)$, that arises as a result of the jump in the value of S when $t = t_0$ may be found from the solution of the Cauchy problem for surface waves with initial condition

$$\frac{\partial h}{\partial t}\Big|_{t=t_0} = \left[\frac{\partial \Phi}{\partial y}\right]_{t_0-0}^{t_0+0}.$$

This problem is equivalent to the problem of wave generation by a pressure pulse. The corresponding analytical solutions are described in the book (Sretensky, 1977).

Note also that the momentum may be carried off by perturbations of other types too. For example, given a density discontinuity (thermocline) near the surface, the formation of a surface wave packet is accompanied by internal wave radiation on thermocline with the wavelength of order envelope length l (Ezersky and Papko, 1986). If the thermocline is sufficiently deep, the surface waves do not interact immediately with the thermocline and its perturbation depends only on the induced flow.

The internal waves also carry off part of the momentum in their propagation. But the total momentum radiated by all long-wave modes in the course of wave packet formation is exactly equal to the quasi-momentum $P_0 = \int_{-\infty}^{+\infty} P_w \, dx$.

1.4 Absorption and scattering of a surface wave packet. Residual vortex

We will now analyse the changes in the mean flow caused by wave dissipation. Consider as an example a two-dimensional train of surface waves. Dissipation of perturbations under the action of viscosity in the boundary layer near the free surface gives rise to a near-surface flow (Phillips, 1977). The density of momentum due to Stokes transport transforms, on wave damping, to the density of momentum of the near-surface flow. This process is localized if the characteristic time of damping is $\tau \ll l/c_g$ where l is a characteristic size of the train and c_g is a group velocity of the wave. Thus, instead of the source and sink system associated with intensity gradients of Stokes transport, a system of near-surface flows is formed having the same mass flux as between the tops and troughs of the wave and the same sources and sinks due to the flow gradients. Consequently, instead of the damped wave packet involved with induced dipole counterflow (Fig. 1.2), a vortex comprising the same dipole counterflow and the near-surface flow that has taken on the quasi-momentum of the wave (Fig. 1.8), remains inside the liquid depth.

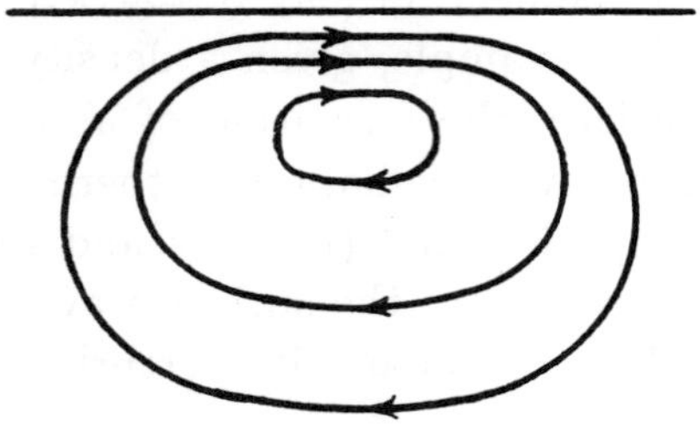

Figure 1.8: Residual vortex.

Note that if the wave packet propagated with group velocity, the residual vortex that emerged on wave damping moves much slower, at a speed that is proportional to the intensity of the near-surface flow. Eventually, viscosity washes out the near-surface, which is accompanied by a decrease of vortex velocity. But the total momentum involved with the residual vortex still remains equal to the total momentum of the wave packet, P_0. An impulse of force P_0 would be needed to excite such a vortex by external force. Besides, exactly the same magnitude of momentum would be carried off by fast long-wave perturbations.

Thus, a residual vortex whose momentum is equal to the wave packet momentum is formed in viscous damping of the surface wave packet. The picture of near-surface sources and sinks for mean flows that is determined by the values of the right-hand sides of system (1.10) remains unchanged so that long waveshaving the scale of the envelope are not radiated in this case.

Wave absorption by a slender freely floating body — a float — occurs in a similar fashion (Fig. 1.9). The absorbed wave packet gives up to the float the momentum due to the Stokes transport (quasi-momentum).

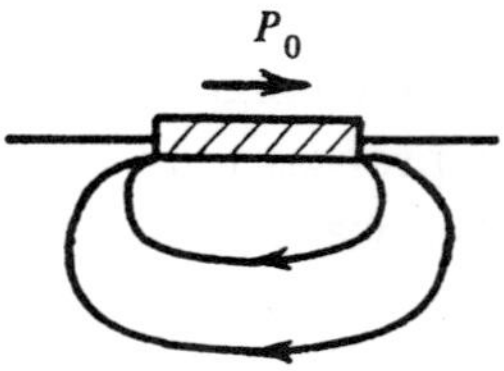

Figure 1.9: Flow around a float.

Consequently, the float starts moving in the direction of the phase velocity of the wave and acquires a momentum P_0. The weight of the float in hydrostatic equilibrium is equal to the weight of the fluid displaced by it. Hence, the float produces the picture of a dipole flow that would be produced by a liquid volume equivalent to the immersed portion of the float and moving at the same speed. Thus the total momentum of the flow is equivalent to the vortex momentum shown in Fig. 1.8, the only difference being that a moving float acts as a near-surface flow. No long-wave perturbations are radiated, like in the case of a moving vortex.

The induced dipole counterflow transforms here to a potential flow around a float. Note that an added mass (Landau and Lifshits, 1986) is a characteristic of a stream-lined body in an incompressible fluid. The added mass is equal to zero only for an infinitely thin body while the total momentum of the system is equal to the momentum P_0 of the float itself.

The absorption of a wave packet by a fixed wave absorber produces a different picture. In this case the sources and sinks involved with the mass flux gradient and, consequently, the induced counterflow disappear. The wave absorber takes on the momentum P_0. Simultaneously, the changes in the field of mass source give rise to radiation of the long-wave perturbations which carry off the momentum $-P_0$. The perturbations escape and the neighborhood of the wave absorber remains in the state of complete rest.

An interesting situation occurs also at wave packet reflection from a motionless obstacle. If the obstacle has the size $h \ll l$, it reflects the wave without immediate effect on mean flow. As a result of wave packet reflection, the dipole that consists of source and sink and generates a potential mean flow changes its sign. Nonstationary sources radiate gravity waves; this process may be investigated within system (1.10). On full reflection, the wave packet momentum changes its sign, consequently, the momentum of the recoil force acting on the reflector equals $2P_0$. At the same time, the long-wave perturbations radiated at reflection must carry off the momentum $-2P_0$.

The scenario is different for the reflection from an infinite vertical wall that serves as a "mirror", not only for surface waves but for mean flow too. In this case, the effect of the reflector may, evidently, be replaced by the interaction with the reflected image of the wave packet in imaginary half-space (Fig. 1.10). Counterpropagating waves generate

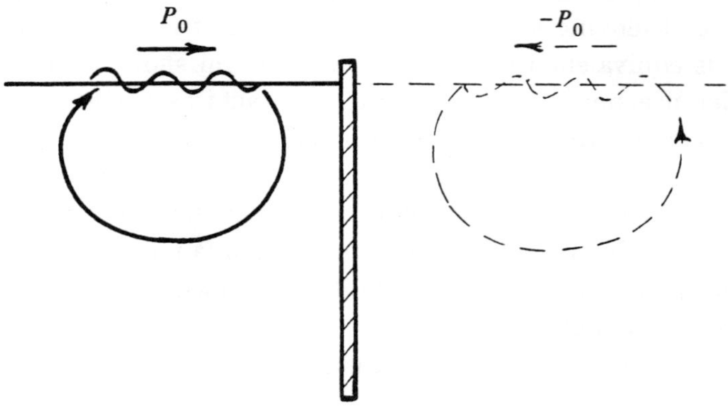

Figure 1.10: Reflection of a wave train at a wall. The dashed curves are for flows in imaginary half-space.

independent sources of a mean field: by averaging the right-hand sides in (1.10) we obtain a sum of intensities of counterpropagating quasi-monochromatic waves. At the same time, equations (1.10) with sources specified in the right-hand sides form a linear system. Thus when counterpropagating wave packets pass through one another their mean flows obey the superposition principle. As a consequence, no long-wave radiation occurs even at reflection from the wall and the total force exerted on the wall by the fluid is equal to zero.

1.5 Ponderomotive force in elastic medium. Induced flows for internal gravity waves

In the previous sections, we considered the motions induced by surface waves in an ideal incompressible fluid. The respective picture of the flow may be quite different for other types of waves in a continuous medium. For instance, a ponderomotive force acting on the particles of the medium at the leading and trailing slopes of the wave packet may emerge for waves in elastic medium. This force acting in a solid gives rise not to a free flow, as was the case with a homogeneous fluid, but instead to a finite particle displacement that is proportional to wave intensity. No residual vortex is formed in wave dissipation; instead, long-wave acoustic perturbations are radiated (Peierls, 1975). Another interesting potentiality that was mentioned in the paper cited above is Cherenkov radiation of sound by a wave packet, given that its group velocity is higher than the speed of light.

One more, an intermediate, case is induced flows of internal gravity waves (IGW) in a stratified medium. An induced vortex flow is formed in the horizontal plane while vertical motion occurs like in an elastic medium: the buoyancy forces compensate the action of ponderomotive forces, which gives rise to a finite vertical displacement of particles in the fluid that is proportional to wave intensity.

Consider equations of dynamics for mean fields of horizontal velocity U and pressure $\overline{p}$ induced by a quasi-monochromatic IGW in which all the variables areproportional to $\exp(-i\omega t + ik_x x + ik_y y + iqz)$:

$$\begin{aligned} \frac{\partial U}{\partial t} + \frac{\nabla_\perp \bar{p}}{\rho_0} + (\mathbf{c}_g \nabla_\perp)\mathbf{P}_{w\perp} &= 0, \\ \nabla_\perp \mathbf{U} &= 0, \end{aligned} \tag{1.26}$$

where $\nabla_\perp = (\partial/\partial x, \partial/\partial y)$, $\mathbf{P}_{w\perp} = \mathcal{E}\mathbf{k}_\perp/\omega$ is the horizontal component of the wave momentum, $\mathbf{k}_\perp = (k_x, k_y)$, $\mathbf{c}_g$ stands for the group velocity of IGWs and

$$\mathcal{E} = \frac{\rho_0}{2}\left(\langle v\rangle^2 + \frac{\langle \sigma^2 \rangle}{N^2}\right)$$

for their energy density expressed through the amplitudes of velocity oscillations and of buoyancy force $\sigma = g\rho/\rho_0$ (ρ being the amplitude of density oscillations and ρ_0, the unperturbed density of the stratified medium), and N denotes Brunt–Väisälä frequency (Gossard and Hooke, 1975).

Equations (1.26) were derived in the paper (Bretherton, 1969) in a quasi-static approximation, when the IGW phase velocity of the moving wave train was small compared to the phase velocity of long-wave disturbances whose scale is equal to that of the train envelope. In this case, the horizontal mean flows are described by a closed system (1.26) and the average buoyancy force $\langle\sigma\rangle$ and vertical displacement $\langle\zeta\rangle$ of particles in the fluid are expressed through the solution of this system:

$$\begin{aligned} \rho_0 <\sigma> &= -\frac{\partial \bar{p}}{\partial z} + (\mathbf{c}_g \nabla) P_{wz}, \\ <\xi> &= \frac{<\sigma>}{N^2}, \end{aligned} \tag{1.27}$$

where $P_{wz} = \mathcal{E}q/\omega$ is a vertical component of the IGW wave momentum.

Horizontal mean flow generation by an IGW train may be considered similarly to surface waves. Towards this end, weintroduce a new variable for mean velocity:

$$\mathbf{V} = \mathbf{U} - \mathbf{P}_{w\perp}/\rho_0 \tag{1.28}$$

Following the law of conservation of wave momentum:

$$\frac{\partial \mathbf{P}_\perp}{\partial t} + (\mathbf{c}_g \nabla)\mathbf{P}_{w\perp} = 0, \tag{1.29}$$

system (1.26) transforms to the form

$$\begin{aligned} \frac{\partial \mathbf{V}}{\partial t} + \frac{\nabla_\perp \bar{p}}{\rho_0} &= 0, \\ \nabla_\perp \mathbf{V} &= Q, \end{aligned} \tag{1.30}$$

where $Q = -\nabla_\perp \mathbf{P}_{w\perp}$.

Thus, according to (1.28), a mean flow can be represented as a superposition of two flows. The first of them possesses the velocity $\mathbf{V}_w = \mathbf{P}_{w\perp}/\rho_0$ and provides local mass transport equal to the wave momentum density (Stokes transport in the case of surface waves). The second flow having the velocity $\mathbf{V}$ is a potential flow produced by source distribution $Q(\mathbf{r}_\perp)$. This flow corresponds to the induced counterflow that was considered above for surface waves and may also be found by means of an electrostatic analogy as a solution of an electrostatic problem of the field of a given charge distribution.

Lines of force for the flows that have the velocities $\mathbf{V}_w$ and $\mathbf{V}$ and are induced by a wave train of rectangular cross-section, as well as a net flow moving with the velocity $\mathbf{U} = \mathbf{V} + \mathbf{V}_w$, are depicted in Fig. 1.11. The net flow contains no mass sources or sinks but is a vortex one. The

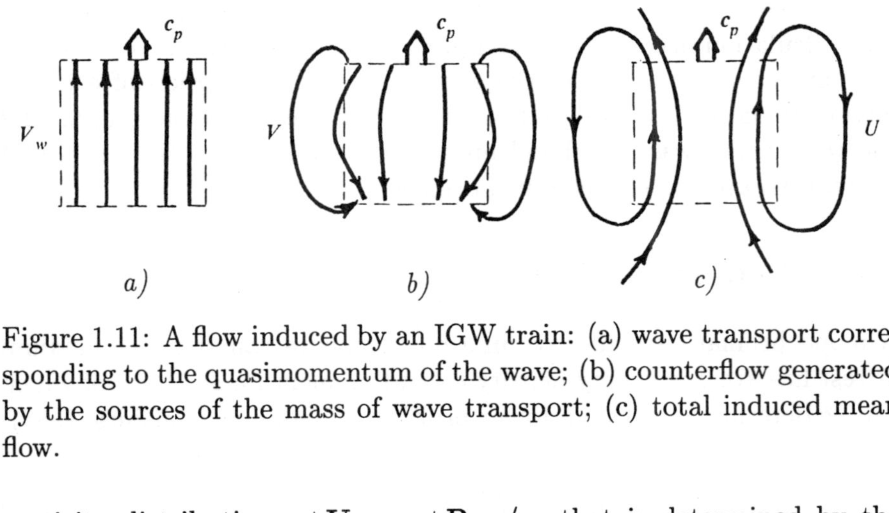

Figure 1.11: A flow induced by an IGW train: (a) wave transport corresponding to the quasimomentum of the wave; (b) counterflow generated by the sources of the mass of wave transport; (c) total induced mean flow.

vorticity distribution $\operatorname{rot}\mathbf{U} = \operatorname{rot}\mathbf{P}_{w\perp}/\rho_0$ that is determined by the shape of the train envelope specifies a dipole picture of lines of force in the form of "spectacles" typical for a vortex pair. However, in contrast to a free vortex pair, the wave-packet induced flow moves with IGW group velocity, accompanying the wave. In the presence of dissipation, the wave gives up its quasi-momentum to the medium, which, like in the case of surface waves, gives rise to a residual vortex whose momentum is equal to the total momentum of the wave packet.

The IGW train propagation is accompanied by mean displacement of level lines, $\langle\zeta\rangle$, that may be found from the given IGW envelope employing (1.27) and by knowing the mean flow induced horizontally. We are concerned here with a quasi-static displacement: it is determined by the local wave intensity and vanishes as soon as the wave packet passes through the given point.

The generation of quasi-monochromatic IGW gives rise to the radiation of long-wave perturbations, i. e., internal waves whose characteristic period is equal to the time in which the wave train envelope changes. These perturbations, like the respective perturbations for surface waves, take on the wave train momentum and carry it off "to the infinity". Generation of long IGW cannot be calculated by means of Eqs. (1.27) and goes beyond a quasi-static approximation, similarly to the flow momentum in a homogeneous fluid (see Sect.1.2) when correct calculation needs to take compressibility into account. The field of the long waves emerging in the generation and absorption of a quasi-monochromatic IGW, as well as the vortex mean flow accompanying the wave train, was calculated numerically by Zhu and Holton (1987) for real parameters of mean atmosphere.

1.6 Acoustic wind

A monochromatic sound wave in a homogeneous compressible medium at rest has average energy density $\mathcal{E}_0$ and momentum $\mathbf{P}_w$ both of which can be determined within a linear problem (Landau and Lifshits, 1986):

$$\begin{aligned} \mathcal{E}_0 &= \frac{1}{2}\left(\rho_0|v^2| + \frac{|p^2|}{\rho_0 c_s^2}\right), \\ \mathbf{P}_w &= \frac{1}{2}\mathrm{Re}\,(\rho\mathbf{v}^*) = \frac{\mathbf{k}}{\omega}\mathcal{E}_0, \end{aligned} \tag{1.31}$$

Here ρ and $\mathbf{v}$ are the amplitudes of density and velocity oscillations calculated in the linear approximation.

The generation and absorption of sound waves in the presence of space–time inhomogeneity of an acoustic field give rise, like in the case of waves on a water surface, to induced mean flows which also possess definite energy and momentum. These flows are described by equations of dynamics of an ideal gas averaged over wave oscillations:

$$\begin{aligned} \frac{\partial \mathbf{U}}{\partial t} + (\mathbf{U}\nabla)\mathbf{U} + \frac{\nabla \bar{p}}{\bar{\rho}} &= 0, \\ \frac{\partial \rho}{\partial t} + \nabla(\bar{\rho}\mathbf{U}) &= Q_m, \end{aligned} \tag{1.32}$$

where $\mathbf{U}$, $\overline{p}$, and $\overline{\rho}$ are mean velocity, density and pressure, respectively, and the mass source, Q_m, is determined by the density gradient of the wave momentum:

$$Q = -\nabla(\rho \mathbf{v}) = -\nabla \mathbf{P}_w \tag{1.33}$$

Equation (1.33) was derived in the approximation of a linear dependence $p(\rho) = c_s^2 \rho$. Allowance for nonlinearity in the equation of state would lead to the appearance in averaged Euler equations of an additional mean force that is analogous to striction force for electromagnetic waves in dielectric. This mean force pushes the gas out of the region occupied by the acoustic field (Landau and Lifshits, 1986).

Similarly to surface waves on water, the mean flows induced by an acoustic wave emerge in the presence of mass sources and sinks produced by inhomogeneous Stokes transport. One can readily show that the Stokes mass transport and the related Lagrange particle drift accompany the propagation of the monochromatic acoustic wave in gas. Assuming the mean Euler velocity to be $U = 0$, we can express the mass flux, like in the case of surface waves, through mean Lagrangian velocity $\langle V_L \rangle$:

$$\mathbf{S}_m = \rho_0 \langle \mathbf{V}_L \rangle = \frac{\rho_0}{2}([\boldsymbol{\xi}^* \times i\mathbf{k}] \cdot \mathbf{v}) = \frac{\rho |v|^2}{2} \frac{\mathbf{k}}{\omega} = \mathbf{P}_w \tag{1.34}$$

The induced flow produced by the source distribution Q_m compensates this mass transport. Note that the Lagrange particle drift in a homogeneous plane wave and the consequent wave-induced Euler counterflow exist only in a compressible medium, in the presence of a longitudinal (with respect to the wave vector) component of velocity fluctuations in the wave. The effect of generation of an induced counterflow by vertically propagating acousto-gravity waves was considered in the paper (Coy et al., 1986) for the explanation of vertical motions observed by means of the Dopler radar technique in mean atmosphere.

An induced flow, similarly to other types of waves, is nonlocally associated with the wave field; it depends, in particular, on boundary conditions. An example is the Euler counterflow that is formed in an ideal fluid in the radiation of an acoustic wave by a piston when the total mass flux is equal to zero (Ostrovsky, 1968). In alternative cases, the total momentum of the medium in a stationary acoustic field may take on different values. Ambiguity in second-order magnitudes with respect to amplitude for nonlinear waves is one of the causes of diverse definitions of energy and momentum of sound in a moving medium (Zarembo and Krasil'nikov, 1966).

Nonstationary processes in sound-induced flows are more complicated than the processes considered above for surface waves. Because of the absence of dispersion for acoustic waves, the large-scale perturbations which take on the momentum in the course of acoustic train generation move with the velocity equal to the wave velocity, i. e., to the speed of sound. The perturbations having the scale of the envelope which accompany the wave train will resonantly increase, drawing energy and momentum from the detected wave. In this case, it is impossible to find a solution similar to the one considered above for surface waves, i. e., a stationary flow induced by a given wave train. Stationary solutions to (1.32) corresponding to mean flows of finite velocity are possible for a wave packet only when sound propagates with dispersion: for higher modes in waveguides, in multiphase media (suspension, fluid with gas bubbles), etc.

Finally, we would like to mention still another type of mean flows accompanying the propagation of acoustic waves. We mean the acoustic flows which are related to the absorption of an acoustic wave. In the presence of damping, part of the acoustic wave momentum is given up to the medium. The resulting bulk force generates a mean flow – acoustic wind. The flows induced by dissipating acoustic beams in a viscous medium (at small Reynolds numbers) were investigated in ample detail both theoretically and experimentally (Eckhart, 1948; Ivanovsky, 1959; Ostrovsky and Papilova, 1974; Rudenko and Soluyan, 1975).

1.7 How to determine wave energy and momentum?

Summing up the wave problems considered above, we would like to emphasize again that there are different ways to determine the energy and momentum density of waves in a medium, depending on which part of energy or momentum of the wave motion is referred to wave-induced mean fields. In particular, different definitions of wave momentum in a medium at rest give, on Galilean transformation, different values of energy density of the wave in a moving medium.

It is far from being trivial to divide the momentum into wave and induced flow components for purely physical considerations in a particular problem because the momentum density has different distributions in different wave systems. The best known example that illustrates this ambiguity is, evidently, the long-known problem of dividing the energy-momentum tensor of electromagnetic waves into the wave part (momentum and energy of the field) and momentum (energy) of the medium (Ginzburg and Ugarov, 1976). According to Abraham, in particular, the wave momentum is related solely to its electric **E** and magnetic **H** strengths:

$$\mathbf{P}^A = \frac{[\mathbf{E} \times \mathbf{H}]}{4\pi c}.$$

The wave propagation in the dielectric is then accompanied by the emergence of a bulk force

$$\mathbf{f}^A = \frac{1}{4\pi c}\frac{\partial}{\partial t}\left([\mathbf{D} \times \mathbf{B}] - [\mathbf{E} \times \mathbf{H}]\right),$$

(where **D** and **B** are the vectors of electric and magnetic fields) acting on the medium, which together with the Lorentz force and striction forces produce mean flows (in liquid) or stresses (in elastic medium) accompanying the wave train. While Minkowski refers the part of the wave-induced momentum caused by the force $\mathbf{f}^A$ to the wave momentum the value of which is

$$\mathbf{P}^M = \frac{[\mathbf{D} \times \mathbf{B}]}{4\pi c}.$$

The latter expression corresponds to the momentum that is lost by the emitter exciting the wave in a dielectric medium. One cannot choose the energy-momentum tensor following Abraham or Minkowski by proceeding only from the laws of conservation of energy and momentum; the choice is usually made for the sake of convenience in solving a concrete problem. In our concept we follow Minkowski.

Summarising this concept we would like to note that the value of quasi-momentum expressed through the wave action (1.3) is taken as a wave momentum. We must take into consideration, however, the wave-induced mean field (flow) which, together with the wave field, make an entity like, for instance, the one shown in Fig. 1.2. This nonlinear object moving with a group velocity of the wave is a generalization of a linear wave packet. The momentum density associated with this object may be divided into two components: a quasi-momentum that is locally proportional to the wave intensity and the density of momentum of induced mean flow that may have an intricate distribution in space. The induced flow momentum in an incompressible fluid is concentrated "at the infinity", i. e., the magnitude of the total momentum is determined by the volume integral of the corresponding momentum density that diverges in an infinitely far region. The magnitude of the total flow momentum is determined uniquely if the large-scale perturbations which carry off a certain momentum in the generation and absorption of the wave packet are taken into consideration. With these perturbations taken into account, the momentum of induced flows is always equal to zero and the total wave momentum is equal to the quasi-momentum.

Such a definition of momentum for a wave train in an incompressible medium follows from comparison with a classical problem of a potential flow around a solid body and proceeds from the impulse of force needed for the excitation of the wave. This definition is most convenient in a theory of wave radiation and absorption in a medium as well as in a theory of wave propagation and scattering in hydrodynamic flows. Besides, we find it highly important that this concept is in conformity with the general theory of oscillations and waves, which allows one to consider various wave problems from a common viewpoint, using a unified physical language.

1.8 Negative energy waves in hydrodynamics

Energy considerations are of primary significance in the investigation of various problems of wave generation and absorption in shear flows. Linear problems are, apparently, concerned with the quasi-energy of a monochromatic wave. We have considered above the physical meaning of the quasi-energy and its relation to the total energy of the system, including wave-induced flows. We have also mentioned that the wave energy may change its sign and become negative. As negative energy waves (NEW) are excited, the total energy of the medium with a wave diminishes. So, no energy is needed to increase the NEW intensity; instead, it should be withdrawn from the system. Waves having energies of different signs in shear flows enable researchers to explain, in particular, various types of instabilities in hydrodynamics. Of course, NEWs may exist only in nonequilibrium media — in flow media, in particular. Examples of such media are, besides hydrodynamic flows, electron beams or beams of charged particles in a plasma.

The concept of negative energy waves was proposed by L.Chua as applied to waves in electron beams back in 1951 (Pierce, 1974) and has since been broadly used in plasma physics and electronics (Kadomtsev *et al.*, 1964; Mikhaĭlovskiĭ, 1974; Nezlin, 1976), while in hydrodynamics, that concept did not arouse interest for a rather long time. Mention should be made, however, of the pioneer paper (Benjamin, 1963) where a classification of hydrodynamic instabilities was presented within which NEWs had a significant role (in particular, Tollmien–Schlichting waves in a boundary layer were referred to NEWs).

It is convenient to illustrate this classification by means of an oscillator equation where the coefficients may have arbitrary signs:

$$m\ddot{x} + \nu\dot{x} + kx = 0, \tag{1.35}$$

where x is the displacement, m stands for mass, ν for friction coefficient, and k is the elasticity of the spring for a mechanical oscillator (Fig. 1.12*a*); or x is the capacitor charge, m the inductance, ν the ohmic resistance in the circuit, and $1/k$ denotes capacitance if we are concerned with electric oscillator (Fig. 1.12*b*).

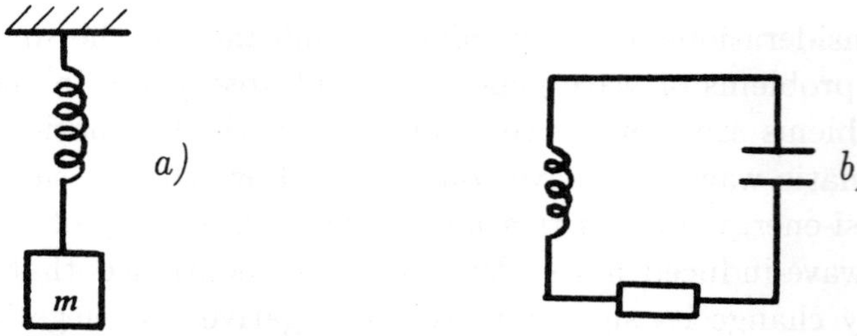

Figure 1.12: Models of mechanical (a) and electrical (b) oscillators.

Multiplying (1.35) by $\dot{x}$ yields an equation of energy balance:

$$\frac{d}{dt}\left[\frac{m(\dot{x})^2 + kx^2}{2}\right] = -2\nu(\dot{x})^2 \tag{1.36}$$

The expression in square brackets,

$$E = \frac{1}{2}\left[m(\dot{x})^2 + kx^2\right],$$

is the total energy of the oscillator. In an equilibrium medium when $E > 0$ and $\nu > 0$, Eq. (1.36) indicates that the energy in the system decreases ($\dot{E} < 0$) due to dissipation. Therefore, in order, for example, to swing a pendulum, definite positive work must be done which then goes over to heat. If $E > 0$ again, but $\nu < 0$, then even small fluctuations may initiate spontaneous building-up of oscillations due to "negative friction" or negative resistance in the LC-circuit (Fig. 1.12*b*). A tunnel diode on the drooping slope of a C-V characteristic may have such a negative resistance. The medium is active in this case, i. e., it may release energy due to nonconservative processes. When $\nu > 0$ but $E < 0$, then, like in the previous case, small fluctuations will increase in time without external energy consumption. In this case, however, the instability is involved with negative energy, E, of the system (which is also a consequence of the system's nonequilibrium) that is conserved in the absence of dissipation. An ordinary "positive dissipation" gives rise to building-up of oscillations and to the growth of the absolute value of E while "negative friction" ($\nu < 0$) favours the damping of oscillations. It is hard to realize such an oscillator in mechanics but it can readily be implemented in electrical engineering (Kashkaryan and Frumkin, 1975, 1976).

Finally, to make our consideration more complete we must mention the case when m and k have opposite signs. Then, one can easily see from (1.35) that the solution does not oscillate independent of the sign of ν and there always exists a single solution growing exponentially with time. This instability occurs in systems with negative mass (Petelin, 1985). It is interesting that the total energy of the system is, in this case, conserved in the absence of dissipation but the energy is transferred from the potential component over to the kinetic one or vice versa.

Let us now come back to waves in a continuous medium and consider the dependence of the signs of wave energy and dissipation on the reference system. Proceeding from the Galilean transformation (1.4) one can see that the wave energy will have opposite signs in motionless and moving media if the phase velocity changes its sign with the change of the reference system, i. e., $\omega - kU < 0$. Alternatively, NEWs may be realized in nonequilibrium media by changing the sign of the quantity $\partial D_0/\partial\omega$ (and, consequently, of the adiabatic invariant I), which is apparent from the relation (1.2). This may occur at the points of the dispersion curve at which the group velocity

$$c_g = \frac{\partial D_0/\partial k}{\partial D_0/\partial \omega} \to \infty.$$

A typical portion of the dispersion curve corresponding to NEWs is depicted in Fig. 1.13.

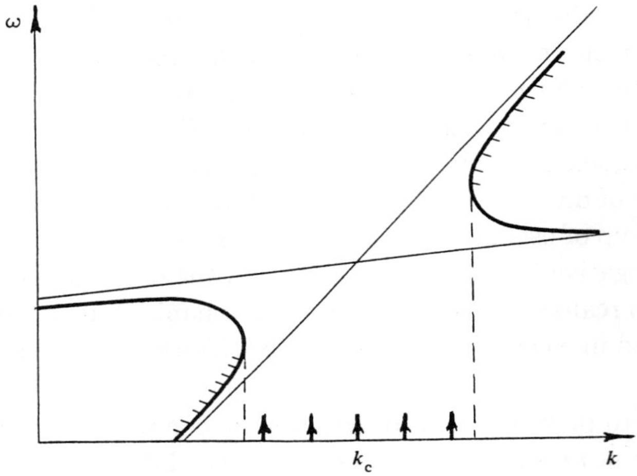

Figure 1.13: Dispersion curves for two interacting waves with different signs of energy. The instability region near $k = k_c$ is marked by arrows. The portions of dispersion curves corresponding to NEWs are hatched.

It has already been said that introducing positive dissipationin the conservative system with NEWs gives rise to the increase of the NEWs in time. The sign of energy depends on the choice of the reference system. However, the dissipative instability *per se*, evidently, does not depend on the reference system. Thus, in passing over to the moving reference system, the dissipation must change its sign together with the wave energy.

The instability in the presence of NEWs may occur not only due to the energy withdrawn from these waves through dissipative processes but also because of the coupling of these waves with other waves having positive energy. We will consider this problem briefly following the paper (Cairns, 1979).

Let the ω, k-plane have two branches of the dispersion curves corresponding to two different waves that are not coupled to one another and are described by the equations $D_1(\omega, k) = 0$ and $D_2(\omega, k) = 0$.

Assume that, at a certain value of k_1, the curves are almost tangent to one another or even intersect so that the first dispersion equality is met at $\omega = \omega_1$ and the second, at $\omega = \omega_2 = \omega_1 + \delta$, where $|\delta| \ll |\omega_1|$. Introducing small coupling between the waves yields, instead of two independent dispersion dependences, a single dispersion relation:

$$D_1(\omega, k)D_2(\omega, k) = \varepsilon \tag{1.37}$$

Sufficiently weak coupling at $k = k_1$ results in the change of the intrinsic frequencies of the waves which take on the form $\omega_1 + \Delta_1$ and $\omega_2 + \Delta_2 = \omega_1 + \delta + \Delta_2$, where $|\Delta_{1,2}| \ll |\omega_1|$. The values of $\Delta_{1,2}$ may be found from an approximate solution of the dispersion equation (1.37) by expanding the $D_{1,2}$ functionsin Taylor series in the vicinity of the point ω_1, k_1:

$$\left[D_1(\omega_1, k_1) + \left[\frac{\partial D_1(\omega_1, k_1)}{\partial \omega}\right] \Delta + \ldots\right]$$

$$\times \left[D_2(\omega_2, k_1) + \left[\frac{\partial D_2(\omega_2, k_1)}{\partial \omega}\right] (\delta + \Delta) + \ldots\right] = \varepsilon \tag{1.38}$$

From this follows a quadratic equation

$$\Delta^2 - \delta\Delta - \varepsilon(\partial D_1/\partial\omega)^{-1}(\partial D_2/\partial\omega)^{-1} = 0 \tag{1.39}$$

with the roots

$$\Delta_{1,2} = \frac{\delta}{2} \pm \sqrt{\delta^2/4 + \varepsilon(k_1)(\partial D_1/\partial\omega)^{-1}(\partial D_2/\partial\omega)^{-1}}$$

The instability occurs if $\Delta_{1,2}$ are complex values, i. e., if

$$\begin{aligned} &\varepsilon(k_1)(\partial D_1/\partial\omega)^{-1}(\partial D_2/\partial\omega)^{-1} < 0, \\ &\delta^2 < -4\varepsilon(\partial D_1/\partial\omega)^{-1}(\partial D_2/\partial\omega)^{-1} \end{aligned} \tag{1.40}$$

From the expressions (1.2) and (1.3) follows that the quantity $\partial D/\partial\omega$ determines the sign of the energy for the wave having the amplitude a. Consequently, the condition of instability may be represented in the form

$$\delta^2 < -\frac{\varepsilon\omega^2|a_1|^2|a_2|^2}{4\mathcal{E}_1\mathcal{E}_2} \tag{1.41}$$

Thus, the instability occurs at sufficiently strong coupling (that is determined by the parameter ε) or, to put it in other words, at small deviation of δ from the exact resonance between two waves having different signs of energy (see Fig. 1.13).

Qualitatively new effects which are absent in equilibrium media appear in nonlinear processes in which NEWs participate (Rabinovich and Fabrikant, 1976). The most illustrious of them is explosive instability which arises in the course of nonlinear interaction of three waves when a high-frequency NEW decays into low-frequency waves of positive energy. The sum of energies of the interacting waves remains constant but the amplitudes of the three modes grow without restriction and turn to infinity in some finite time (Rabinovich and Reutov, 1973). The explosive instability proved to occur in different hydrodynamic flows which was shown in (Voronovich and Rybak, 1978; Moiseev et al., 1983; Craik, 1985).

1.9 The relationship between kinetic and potential energies of linear perturbations on a vortex sheet

It is common knowledge that average potential and kinetic energies are equal in both surface and internal waves of small amplitude in a motionless fluid (Phillips, 1977) by virtue of the virial theorem (Landau and Lifshits, 1973b) for arbitrary mechanical systems. An average value of the Lagrangian function (Whitham, 1974) that coincides with the difference between kinetic and potential energies is equal to zero in this case. However, average kinetic and potential energies do not necessarily coincide in nonequilibrium media such as, for example, stratified fluids with shear flows. Actually, the kinetic energy density in a shear flow is proportional to

$$\int_{-\infty}^{\infty} \rho(z)[(\mathbf{U}+\mathbf{v})^2 - \mathbf{U}^2]\,dz = \int_{-\infty}^{\infty} \rho(z)[\mathbf{v}^2 + 2(\mathbf{U}\cdot\mathbf{v})]\,dz, \qquad (1.42)$$

where **U** is the velocity of external flow and **v**is the perturbation of fluid velocity which depends explicitly on **U** while the potential energy is **U**-independent and is determined solely by the displacement of isopycnic surfaces. In this case, the Lagrangian function is no longer determined by the difference between kinetic and potential energies although the average value of Lagrangian persists to be equal to zero.[1] Such systems are well known in mechanics and are referred to as non-natural systems (Gantmacher, 1966). The second term,

$$2\int_{-\infty}^{\infty} \rho(z)(\mathbf{U}\cdot\mathbf{v})\,dz,$$

in the expression for kinetic energy indicates that its sign may be changed if the velocity oscillations are in counterphase with the oscillations of the isopycnals (for details see the text below). This may lead, in conformity with what was said in Sect. 1.8, to negative total energy of wave perturbations if the flow velocities are large enough. It is interesting to investigate the relation between different types of energies with the variation of flow velocity and for different wave numbers using a unified approach. As a simple example let us analyse the average kinetic–potential energy ratio for small perturbations on a vortex sheet in a fluid that consists of two layers of different densities, with surface tension forces between the layers taken into account.

Consider a classical Kelvin–Helmholtz model of an ideal fluid that consists of two infinitely deep layers with the surface tension coefficient σ between layers. Suppose that the upper, lighter fluid of density ρ_1 moves at a constant speed U relative to the motionless lower fluid of density ρ_2. Introducing a coordinate system with the z-axis directed vertically upwards and the x-axis directed along the layer interface one can write governing equations of hydrodynamics and boundary conditions in a linear approximation, assuming the motion to be two-dimensional:

$$\Delta\varphi_1 = 0,\quad z > 0; \qquad \Delta\varphi_2 = 0,\quad z < 0;$$

[1]The latter is valid because the Lagrangian of linear perturbations is proportional to the function $D(\omega, k)$ from which, given it is qual to zero, follows a dispersion equation.

$$\frac{\partial \eta}{\partial t} + U\frac{\partial \eta}{\partial x} = \frac{\partial \varphi_1}{\partial z}, \qquad \frac{\partial \eta}{\partial t} = \frac{\partial \varphi_2}{\partial z}, \quad z = 0; \tag{1.43}$$

$$\rho_1 \left(\frac{\partial \varphi_1}{\partial t} + U\frac{\partial \varphi_1}{\partial x} + g\eta \right) = \rho_2 \left(\frac{\partial \varphi_2}{\partial t} + g\eta - \sigma \frac{\partial^2 \eta}{\partial x^2} \right), \quad z = 0$$

Here, $\varphi_{1,2}$ are the velocity potentials of the upper and lower fluids, respectively, so that $v_1 = Ux_0 + \nabla\varphi_1$ for $z > 0$ (x_0 is a unit vector along the x-axis) and $v_2 = \nabla\varphi_2$; $\eta(x,t)$ is the deviation of the interface from the unperturbed level $z = 0$. Solution of the system (1.43) has the form:

$$\begin{aligned} \eta(x,t) &= \eta_0 e^{i(\omega t - kx)}, \\ \varphi_1(x,z,t) &= -i\left(\frac{\omega}{k} - U\right) e^{-kz}\eta(x,t), \\ \varphi_2(x,z,t) &= i\frac{\omega}{k} e^{kz}\eta(x,t) \end{aligned} \tag{1.44}$$

with the frequency ω related to the wave number k by the following dispersion equation ($a = \rho_1/\rho_2$):

$$\omega_{1,2} = \frac{akU \pm \sqrt{(1-a^2)gk + (1+a)\sigma k^3/\rho_2 - ak^2U^2}}{1+a} \tag{1.45}$$

Following Phillips (1977) we can write the kinetic energy of perturbations averaged over wave period:

$$T = \left\langle \frac{\rho_2}{2} \int_{-\infty}^{\eta} (\nabla\varphi_2)^2 \, dz \right\rangle + \left\langle \frac{\rho_1}{2} \int_{\eta}^{\infty} [(Ux_0 + \nabla\varphi_1)^2 - U^2] \, dz \right\rangle$$

$$\approx \left\langle \frac{\rho_2}{2} \int_{-\infty}^{0} (\nabla\varphi_2)^2 \, dz \right\rangle + \left\langle \frac{\rho_1}{2} \int_{\eta}^{0} 2U\frac{\partial \varphi_1}{\partial x} \, dz \right\rangle + \left\langle \frac{\rho_1}{2} \int_{0}^{\infty} (\nabla\varphi_1)^2 \, dz \right\rangle \tag{1.46}$$

The angular brackets in (1.46) denote time averaging and the approximation used here corresponds to the change of the limits of integration in the first and third terms in the right-hand side from $\eta(x,t)$ to zero, which is consistent with the quadraticapproximation for T. Note the important contribution of the second term in the right-hand side of (1.46). This term is responsible for negative kinetic energy if the displacement of the interface, η, isin counterphase with the horizontal projection of velocity in the wave $\partial\varphi_1/\partial x$. Substituting (1.44) into (1.46) yields

$$T = \frac{\rho_2}{4k}\left[\omega^2 + a(\omega^2 - U^2k^2)\right]\eta_0^2 \tag{1.47}$$

An average potential energy (Phillips, 1977) is now calculatedto be

$$\begin{aligned} V &= (\rho_2 - \rho_1)g\langle\eta^2\rangle + \sigma\langle(\nabla\eta)^2\rangle \\ &= \frac{\rho_2}{4}\left[(1-a)g + \frac{\sigma k^2}{\rho_2}\right]\eta_0^2 \end{aligned} \tag{1.48}$$

from which follows that the potential energy of perturbations is always positive and does not depend on flow velocity U. We will now write a relation for T/V replacing ω, employing the expression (1.43):

$$\frac{T}{V} = 1 - \frac{2aU}{1+a}\frac{kU \pm \sqrt{(1-a^2)gk + (1+a)\sigma k^3/\rho_2 - ak^2U^2}}{(1-a)g + \sigma k^3/\rho_2} \tag{1.49}$$

Let us now pass over to dimensionless variables: $\varepsilon = 1-a$, $u = U/U_*$, $\kappa = k/k_*$, where

$$U_* = \sqrt[4]{\frac{4\varepsilon g\sigma}{\rho_2}\left(\frac{1+a}{a}\right)^2} \quad \text{and} \quad k_* = \frac{a}{1+a}\frac{\rho_2 U_*^2}{2\sigma}$$

are a critical velocity at which Kelvin–Helmholtz instability occurs and the wave number of the disturbance which starts to grow first when $U = U_*$, respectively. In the new variables we have:

$$\frac{T}{V} = 1 - 4\frac{\kappa u^2 \pm \sqrt{a\kappa(1+\kappa^2-2\kappa u^2)/2}}{1+\kappa^2} \tag{1.50}$$

From (1.50) follows that, for $u = 0$, the average kinetic energyof both modes is equal to the average potential density, as was to be expected.

While for $u \neq 0$, these energies, in a general case, are not equal and approach one another only as $\kappa \to 0$ and $\kappa \to \infty$ because the velocity shift weakly affects the properties of sufficiently long or short waves. The dependences of T/V on κ are plotted in Fig. 1.14 for different values of u and $a \approx 1$. Curve 1 corresponds to $u = 1/\sqrt{2}$ at which the kinetic

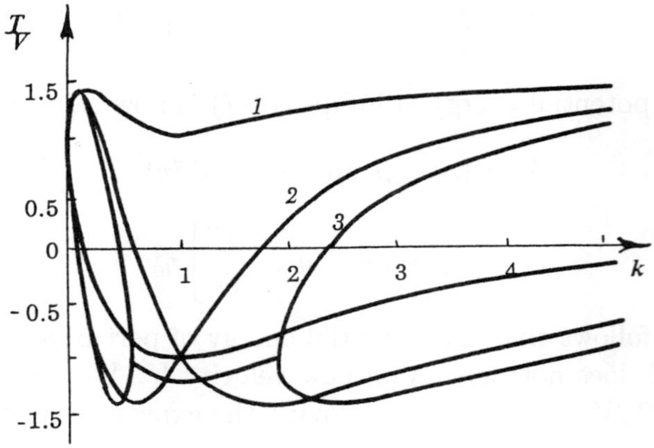

Figure 1.14: The ratio of average kinetic and potential energies T/V versus dimensionless wavenumber κ for waves on a vortex sheet in a liquid with two layers of different densities and with surface tension at the interface. The figures mark the curves corresponding to different velocities of relative motion of the layers.

energy of one of the modes is equal to the potential energy having an opposite sign for $\kappa = 1$, i. e., the total wave energy turns to zero when $\kappa = 1$. At larger values of u, negative energy waves appear in the system considered. The Kelvin–Helmholtz instability occurs starting from $u = 1$, which is shown by curve 2 in Fig. 1.14. Finally, when $u = 1$, frequencies $\omega_{1,2}$ become complex in a definite interval of wave numbers:

$$u^2 - \sqrt{u^4 - 1} \leq \kappa \leq u^2 + \sqrt{u^4 - 1},$$

which indicates that there is Kelvin–Helmholtz instability at these wavelengths. The totalenergy of growing perturbations changes exponentially with time, but the average kinetic-potential energy ratio is finite.

Substituting into (1.50) the real part ω from (1.45), for the growing perturbations we obtain

$$\frac{T}{V} = -\frac{2\kappa u^2}{1+\kappa^2} \tag{1.51}$$

Curve 3 in Fig. 1.14 is plotted for $U = 1.1$. One can easily see that the total energy of perturbations turns to zero ($T/V = -1$) at the point where Kelvin–Helmholtz instability emerges. Maximal excess of the kinetic energy over the potential one outside the region of Kelvin–Helmholtz instability depends on a but its absolute value is never higher than $\sqrt{2}$.

Let us now analyse a shear flow in the absence of surface tension between layers ($\sigma = 0$) which is typical for interlayers in the ocean. It is clear from (1.45) that for any arbitrary small flow velocity there always exists Kelvin–Helmholtz instability in the region of sufficiently large wave numbers

$$k \geq k_0 = \frac{(1+a)g}{aU^2}.$$

Designating $\kappa = k/k_0$, from formula (1.49) ($\sigma = 0$) we obtain expressions for T/V for $\kappa \leq 1$ and $\kappa > 1$:

$$\begin{aligned} \frac{T}{V} &= 1 - 2\left[\kappa \pm \sqrt{a\kappa(1-\kappa)}\right], \quad && \kappa \leq 1; \\ \frac{T}{V} &= -\kappa, && \kappa > 1. \end{aligned} \tag{1.52}$$

The dependences of T/V on κ are presented in Fig. 1.15 for different values of a. Curves 1–3 correspond to $a = 1;\ 0.1;\ 0.01$, respectively. One can easily find from (1.52) that outside the instability region we will have

$$\max\frac{T}{V} = -\min\frac{T}{V} = \sqrt{1+a} \leq \sqrt{2}.$$

These values are attained at

$$\kappa_{1,2} = \frac{1}{2}\left(1 \pm \frac{1}{\sqrt{1+a}}\right).$$

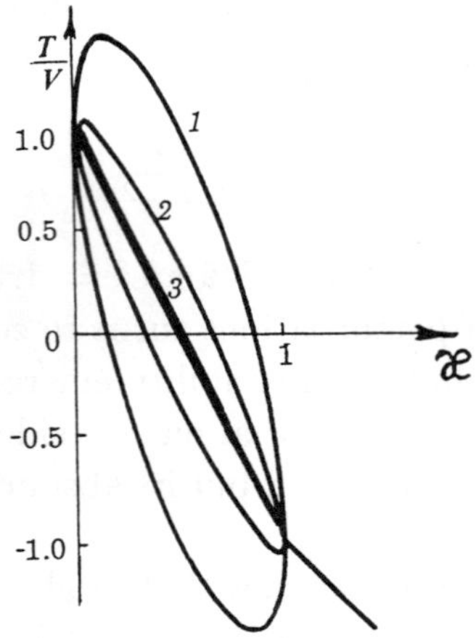

Figure 1.15: Same as in Fig. 1.14 but without account of the surface tension at the interface.

The absolute value of the T/V ratio is always greater than unity in the instability region and increases without restriction as κ grows, but the above expressions hold for the values of κ slightly higher than unity when the ratio of the imaginary part of frequency to the real one is small.

1.10 Spectral evolution of a wave packet

The generation of a mean flow by a quasi-monochromatic wave is a nonlinear effect which may play a significant role in the propagation of nonlinear waves in a medium. In particular, the interaction of the wave packet and the mean flow induced by this packet causes the change of the phase velocity of the wave. The wave self-action arising in this process determines quite a few peculiarities of the nonlinear dynamics of modulated waves. The self-action of a quasi-monochromatic wave is described, to a first approximation in nonlinearity, by a complex parabolic equation (nonlinear Schrödinger equation) for complex amplitude $a(x, t)$

(Whitham, 1974):

$$i\left(\frac{\partial a}{\partial t}+c_g\frac{\partial a}{\partial x}\right)+\kappa\frac{\partial^2 a}{\partial x^2}+\alpha_0 a|a|=0. \tag{1.53}$$

Here, $c_g = \partial\omega/\partial k$ is the group velocity of the wave, the parameter $\kappa = \frac{1}{2}\partial^2\omega/\partial k^2$ describes dispersion, and α stands for the local nonlinear frequency shift that is determined, in particular, by the generation at nonresonance higher harmonics and their inverse effect on the wave. Equation (1.53) is solved by exact methods of the theory of nonlinear waves (Zakharov *et al.*, 1980). Arbitrary, bounded initial perturbation in this equation, when $\alpha\kappa > 0$, decays into envelope solitons, i. e., steadily moving wave packets of the form

$$a = A\exp[-i\alpha A^2 t/2 - i\kappa q^2 t + iq(x-Vt)]\cosh^{-1}[\tilde{k}(x-Vt)] \tag{1.54}$$

(where $\tilde{k} = A(\alpha/2\kappa)^{-1/2}$ and $V = c_g + 2\kappa q$) and nonsoliton oscillating wave trains whose amplitude decreases in time. The solitons are stable and they retain their amplitude and velocity on collisions. The envelope solitons may also appear as a result of modulation instability of a spatially homogeneous solution of Eq. (1.53), given weak dissipation and instability.

The wave interaction with induced mean flow is responsible for additional nonlinearity in the equation for complex amplitude. This nonlinearity is nonlocal, in a general case, which makes the evolution of modulated waves significantly more complicated. In particular, the equations of dynamics of mean flow (1.10) taken in a quasi-static approximation for waves on deep water may be combined with a complex parabolic equation which contains an additional term responsible for a Dopler correction to the wave frequency on mean flow:

$$\begin{gathered}
i\left(\frac{\partial a}{\partial t}+c_g\frac{\partial a}{\partial x}\right)+\kappa\frac{\partial^2 a}{\partial x^2}+\alpha_0 a|a| = s_0 a\frac{\partial\Phi}{\partial x}|_{y=0} \\
\Delta\Phi = 0 \\
\frac{\partial\Phi}{\partial y} = s_1\frac{\partial|a|^2}{\partial x} \quad \text{for} \quad y=0
\end{gathered} \tag{1.55}$$

This two-dimensional boundary value problem was obtained in the paper (Dysthe, 1979). Its solutions, in particular, a modification of modulation instability of a homogeneous wave due to the interaction with mean flow, were also analysed in the paper (Tomita, 1986).

Self-modulation of quasi-monochromatic waves may be investigated in a general form for a nearly one-dimensional mean flow, like it was done in Sect. 1.3, employing a shallow waterapproximation. A general model of the nonlinear self-action of a high-frequency wave, taking into account the interaction of high-frequency waves with a low-frequency mode as well as dissipative effects, has a form

$$\begin{aligned} i\left(\frac{\partial a}{\partial t}+c_g\frac{\partial a}{\partial x}\right)+\kappa\frac{\partial^2 a}{\partial x^2}+\alpha_0 a|a| &= \beta na+i\hat{D}_v(a)+i\hat{N}(a), \\ \frac{\partial^2 n}{\partial t^2}-c^2\frac{\partial^2 n}{\partial x^2} &= \delta\frac{\partial^2|a|^2}{\partial x^2}+\hat{D}_s\left(\frac{\partial n}{\partial t}\right). \end{aligned} \tag{1.56}$$

Here, $n(x,t)$ denotes the perturbation of refractive index for the high-frequency wave, c is the phase velocity of the low-frequency wave (its dispersion is neglected), $\hat{D}_v$ and $\hat{D}_s$ are the linear operators responsible for dissipation and instability of high-frequency and low-frequency waves, respectively, and $\hat{N}$ is the operator of the nonlinear dissipation that restricts the growth of high-frequency waves. System (1.56) describes the dynamics of Langmuir waves in a nonisothermal plasma (Gorev et al., 1976), the surface–internal wave interaction in the ocean (Petrov, 1979), and so on.

In the absence of dissipations ($\hat{D}_v=\hat{D}_s=\hat{N}=0$), the system (1.56), similarly to Eq. (1.53), has a solution in the form of solitons that are long-lived formations here too, and weakly interact with one another. Consider the effect exerted on the solitons by small nonconservative additions such as dissipation and instability (Fabrikant, 1984). Small nonconservative additions weakly affect the shape of the envelope soliton but lead to slow evolution of its parameters: amplitude A and carrier wave number q. The possibility of a narrow range variation of the wave numbers under the action of linear dissipation of particular type was noted for Langmuir waves (Chukbar and Jan'kov, 1977) and for wind waves on deep water (Fabrikant, 1980a). Here, we will consider a broad-range evolution of the spectrum of envelope soliton for a weakly nonlinear wave caused by linear and nonlinear dissipation (instability) with arbitrary spectral dependence. Of particular interest is the adiabatic spectral displacement of solitons in a broad range of wave numbers, because it may give rise to spectral energy transfer and to substantial restructuring of nonlinear oscillations. This phenomenon, in particular, must play a significant role in strong wave turbulence in dissipative systems having a nondecaying spectrum where modulation instability may cause generation of the envelope solitons of quasi-monochromatic waves (Kijashko *et al.*, 1975).

1.10.1 The evolution of Schrödinger solitons

Consider first the evolution of envelope solitons in a single-wave approximation when the amplitude obeys a nonlinear equation:

$$i\left(\frac{\partial a}{\partial t}+c_g\frac{\partial a}{\partial x}\right)+\kappa\frac{\partial^2 a}{\partial x^2}+\alpha a|a|^2=i\mu\hat{R}(a), \tag{1.57}$$

where

$$\mu\hat{R}(a)=\gamma_0 a-i\gamma_1\frac{\partial a}{\partial x}-\gamma_2\frac{\partial^2 a}{\partial x^2}$$

$$-\left[\rho_0 a|a|^2-i\frac{\rho_1}{2}\left(|a|^2\frac{\partial a}{\partial x}-a^2\frac{\partial a^*}{\partial x}\right)+ia\hat{L}\left(|a|^2\right)\right],$$

and μ is a small parameter determining the value of nonconservatism (dissipation or pumping of energy). Equation (1.57) may be derived, in particular, from (1.56) (then $\alpha=\alpha_0+\beta\delta/c^2$) for slow motions having a characteristic velocity $V\ll c$ in a quasi-static approximation, when the term $\partial^2 n/\partial t^2$ may be omitted in (1.56).

Equation (1.57) describes the self-action of a weakly nonlinear narrow-spectrum wave in a medium with weak dispersion. The right-hand side of this equation includes the first terms of power expansion in amplitude and wave number for the dissipation operator in a long-wave approximation. The terms linear with respect to amplitude in the operator $\hat{R}(a)$ are found by expanding the increment $\gamma(k)$ in the operator:

$$\hat{D}_v(a) = \frac{1}{2\pi} \int_{+\infty}^{-\infty} \gamma(k) a_k e^{ikx} \, dk \tag{1.58}$$

(a_k is the Fourier transform of the complex amplitude $a(x)$) in power series of the wave number

$$\gamma_0 = \gamma(k=0), \qquad \gamma_1 = \frac{\partial \gamma}{\partial k}_{k=0}, \qquad \gamma_2 = \frac{1}{2}\left(\frac{\partial^2 \gamma}{\partial k^2}\right)_{k=0} \tag{1.59}$$

The two first nonlinear terms in $\hat{R}(a)$ are analogous to the series expansion of the operator of nonlinear dissipation of the high-frequency wave in cubic approximation with respect to amplitude, with

$$\rho_0 = \rho(k=0), \qquad \rho_1 = \frac{d\rho}{dk}|_{k=0} \, , \tag{1.60}$$

where $\rho(k)$ is the Landau coefficient that is responsible for the saturation of intensity of the high-frequency oscillations in spatially homogeneous regime. Actually, if the solution to (1.57) is taken in the exponential form $a = Ae^{ikx}$ ($k =$ const), one can readily verify that the intensity $|A|^2$ must meet the equation

$$\frac{d}{dt}|A|^2 = 2(\gamma_0 + \gamma_1 k + \gamma_2 k^2)|A|^2 + 2(\rho_0 + \rho_1 k)|A|^4 \tag{1.61}$$

Finally, the last term in $\hat{R}(a)$ describes the nonlocal mechanism of nonlinearity related to dissipation of the mean flow induced by the wave packet. In particular, the linear operator $\hat{L}$ may be expressed in a quasi-static approximation through the operator, $\hat{D}_s$, of dissipation of a low-frequency wave in system (1.56). By expanding $\hat{L}$ in the power series of the wave number we will restrict ourselves to the approximation $\hat{L}(|a|^2) = sd|a|^2/dx$.

It should be noted that Eq. (1.57) takes into account the dependence of nonlinear dissipation on wave number for which the terms proportional to A (the dispersion of nonlineardissipation) are responsible, but neglects the dispersion of conservative nonlinearity, i. e., the terms determined by small imaginary corrections to the coefficients ρ_1 and s are omitted. An account of these corrections may only change the profile of nonlinear waves but it weakly affects the evolution of their characteristic parameters.

Arbitrary bounded perturbation decays into the solitons (1.54) in a nonconservative approximation when $\hat{R} = 0$ in Eq. (1.57). Weak nonconservatism involves slow variations in the soliton amplitude A and in the correction to the carrier wave number q. Evolution equations for these parameters may be derived by means of an asymptotic method (Aranson *et al.*, 1982). In the first approximation relative to μ, these equations may also be obtained from the laws of conservation of the number of particles and of quasi-momentum:

$$I = \int_{-\infty}^{+\infty} |a|^2 dx, \qquad P = \frac{l}{2i} \int_{-\infty}^{+\infty} \left(a^* \frac{\partial a}{\partial x} - a \frac{\partial a^*}{\partial x} \right) dx. \tag{1.62}$$

Note that Eq. (1.57) has an infinite number of first-order integrals for $\hat{R} = 0$. Soliton evolution, however, depends only on the variation of the two first intergals: I and P (Karpman, 1979). The changes in the other first-order integrals in the course of soliton evolution are, in a general case, related to radiation of small oscillatory perturbations, which allows the soliton to conserve its shape.

Consequently, we have

$$\begin{aligned} \frac{dI}{dt} &= 2\mu \mathrm{Re} \int_{-\infty}^{+\infty} a \hat{R}(a) \, dx, \\ \frac{dP}{dt} &= 2\mu \mathrm{Im} \int_{-\infty}^{+\infty} a \frac{\partial}{\partial x} \hat{R}(a) \, dx. \end{aligned} \tag{1.63}$$

Substituting the soliton (1.54) into (1.63) yields in the first-order approximation the following relations

$$I = 2A\sqrt{\frac{2\kappa}{\alpha}}, \qquad P = qI,$$

$$\frac{dI}{dt} = I\left[2(\gamma_0 + \gamma_1 q + \gamma_2 q^2) + \frac{2}{3}\gamma_2 \tilde{k}^2 - \frac{4}{3}(\rho_0 + \rho_1 q)A\right], \tag{1.64}$$

$$\frac{dP}{dt} = qI\left[2(\gamma_0 + \gamma_1 q + \gamma_2 q^2) + \frac{2}{3}\gamma_2 \tilde{k}^2 - \frac{4}{3}(\rho_0 + \rho_1 q)A^2\right] + \frac{4}{3}\gamma_1 \tilde{k} A^2 + \frac{16}{15} s\tilde{k} A^4.$$

Mention should be made that the total wave number $k_w = k + q$ of the high frequency carrier is a characteristic feature of the envelope soliton. The coefficients α, κ, γ, ρ, and s in (1.64) are the functions of the variable k_w and are written as a series expansion in $q = k_w - k$. Substituting the variable

$$a = b\exp[iqx - i(cq + \kappa q^2)t] \tag{1.65}$$

shows that the amplitude b of the wave having the wave number k_w also obeys Eq. (1.57) in which the coefficients that are functions of the wave number are calculated for $k = k_w$. Thus, for the amplitude A of the envelope soliton and for the wave number k_w from (1.64) follow the relations

$$\begin{aligned} \frac{dA}{dt} &= A\left[2\gamma + \frac{A^2}{6}\frac{d(\alpha\kappa)}{dk_w}\frac{d\gamma}{dk_w} + \frac{\alpha A^2}{6\kappa}\frac{d^2\gamma}{dk_w^2} - \frac{4}{3}\rho A^2\right], \\ \frac{dk_w}{dt} &= \frac{\alpha A^2}{3\kappa}\left(\frac{d\gamma}{dk_w} + \frac{4}{5}sA^2\right). \end{aligned} \tag{1.66}$$

to an accuracy of the terms of order A^4. Note that system (1.66) does not contain the parameter k and, consequently, retains its form with relatively large changes in $k_w(|k_w - k| \geq k)$ when the complex amplitude $a(x,t)$ becomes a rapidly oscillating function and Eq. (1.57) is not valid.

Equations (1.66) describe the evolution of one-dimensional envelope solitons in a medium with instability and dissipation. The adiabatic approximation considered above holds only for relatively weak dissipation, when the soliton perturbed by nonconservative corrections reconstructs its shape faster than its parameter changes, i. e., under the condition

$$\gamma \ll \alpha A^2, \quad k\frac{d\gamma}{dk} \ll \alpha A^2, \quad k^2\frac{d^2\gamma}{dk^2} \ll \alpha A^2, \quad \rho \ll \alpha, \quad sk \ll \alpha. \tag{1.67}$$

Now consider different variants of envelope soliton evolution within model (1.66). The solitons decrease in amplitude, simultaneously being broadened in the dissipative medium ($\gamma(k) = -\nu(k) < 0$). While the conditions (1.67) hold, damping occurs in the nonlinear regime which, at low amplitudes, is described by a simplified system

$$\frac{dA}{dt} = -2\nu A, \qquad \frac{dk_w}{dt} = -\frac{\alpha A^2}{3\kappa}\frac{d\nu}{dk_w} \tag{1.68}$$

The damping rate of the soliton amplitude is twice the dampingrate $\nu(k)$ of a monochromatic wave and the wave number changes so that the wave packet spectrum is shifted to the region of smaller values of damping rate. When the soliton having the initial parameters A_0, k_0 damps down to infinitely small amplitudes, the wave number k_w tends to k_∞ (see Fig. 1.16*a*) where k_∞ is found from

$$A_0^2 = -12\nu \int_{k_0}^{k_\infty} \left(\alpha \frac{d\nu}{dk}\right)^{-1} dk$$

Taking into account $\alpha A_0^2/\kappa k_0^2 \ll 1$ we have

$$\frac{k_\infty - k_0}{k_0} = -\frac{\alpha(k_0) A_0^2}{12 k_0 \kappa(k_0) \nu(k_0)} \frac{d\nu}{dk}_{|k=k_0} \tag{1.69}$$

Consider as an example viscous damping of surface gravity waves on deep water (swell). The dimensionless amplitude $a = k\eta$ (where η is the amplitude of surface displacement) of weakly nonlinear two-dimensional gravity waves obeys, in a conservative approximation, Eq. (1.53) where

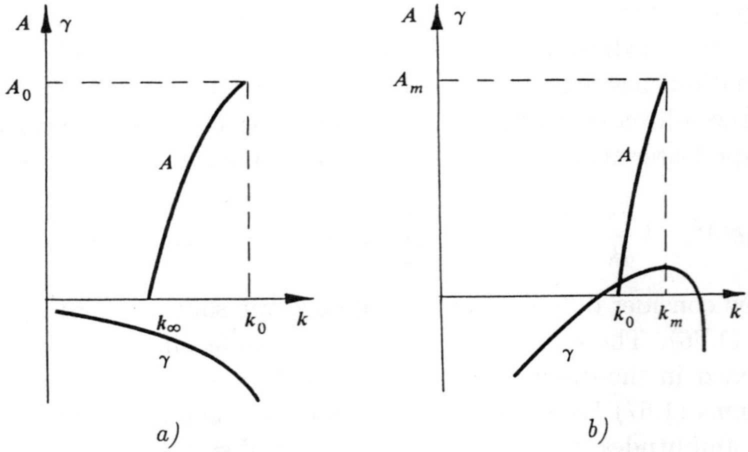

Figure 1.16: Spectral dependence of the growth rate $\gamma(k)$ and variation of the amplitude A and wavenumber k of envelope soliton in a medium with dissipation (a) and instability (b).

$\tilde{R} = 0$, $\alpha = -2(gk)^{1/2}$, and $\kappa = -(1/8)\sqrt{g/k^3}$ (Yuen and Lake, 1982). The development of modulation instability of gravity waves leads to the formation of envelope solitons of the form (1.54) which move along the wave number spectrum, with simultaneous amplitude damping. Substituting into (1.69) the rate of viscous damping, $\nu(k) = 2\nu_w k^2$ (where $\nu_w = 0.01\,\mathrm{cm^2 s^{-1}}$ isthe kinematic viscosity of water) (Landau and Lifshits, 1986), yields for the initial amplitude, $A_0 = 0.2$, a relative displacement of the wave numberin the course of damping:

$$\frac{k_\infty - k_0}{k_0} = -\frac{8}{3}A_0^2 = -0.11. \tag{1.70}$$

It should be emphasized that the nonlinear regime of damping holds down to very small amplitudes. Indeed, the conditions (1.67) are violated only for $A \le A_\nu = (\nu_w^2 k^3/g)^{1/4}$. For the wavelength $\lambda = 1\,\mathrm{m}$ we have $A_\nu = 7 \cdot 10^{-5}$, which corresponds to maximal amplitude of surface displacement $\eta = 10^{-2}\,\mathrm{mm}$.

The soliton amplitude increases in the active medium ($\gamma > 0$) and the wave number shifts towards the increasing increment (Fig. 1.16*b*). A stable equilibrium state with small amplitude is possible in system (1.66) for $k_w = k_m$ when the increment $\gamma(k)$ takes on the maximal value

$$\frac{d\gamma}{dk}\,|_{k+k_m} = 0, \qquad \varepsilon = -\frac{1}{2}\frac{d^2\gamma}{dk^2}\,|_{k+k_m} > 0. \tag{1.71}$$

The approximation $\gamma(k) = \gamma(k_m) - \varepsilon(k - k_m)^2$ holds near this state, and the equilibrium value of amplitude is then found as

$$A = A_m = \sqrt{2\gamma(k_m)}\left[\frac{4}{3}\rho(k_m) + \frac{\alpha(k_m)\varepsilon}{3\kappa(k_m)}\right]^{-1/2}. \tag{1.72}$$

Thus, small amplitude solitons that emerge from an initial perturbation in a weakly nonequilibrium medium equalize their amplitudes and wave numbers and, thus, acquire identical velocities equal to the group velocity of the most unstable harmonic. Further evolution is determined by the interaction of solitons having close parameters which gives rise to a periodic chain of solitons with alternating phases, i. e., a periodic modulation wave is formed (Aranson *et al.*, 1982).

As the amplitude increases, the nonlocal interaction with mean flow for which the coefficient $s(k)$ is responsible in (1.66) must be taken into account. The wave number in this case deviates from $k = k_m$.

1.10.2 Mean flow dissipation. Fast envelope solitons

Solitons of the form (1.54) are also solutions to system (1.56) in a conservative approximation ($\hat{D}_v = \hat{D}_s = \hat{N} = 0$) if the parameter α in (1.54) is replaced by $\alpha = \alpha_0 + \beta\delta/(c^2 - v^2)$. Solitons of this type are formed, for example, as a result of developing modulation instability of Langmuir waves in nonisothermal plasma. Much attention was paid to the investigation of their properties in an attempt to construct a theory of strong plasma turbulence (Gorev *et al.*, 1976).

Let us consider their evolution under the action of the dissipation of high-frequency and low-frequency modes. Towardsthis end, like in the previous section, we will calculate the velocity of variation of the number of quasi-particles and quasi-momentum:

$$I = \int_{-\infty}^{+\infty} |a|^2 \, dx, \qquad P = \frac{1}{2i} \int_{-\infty}^{+\infty} \left[a^* \frac{\partial a}{\partial x} - a \frac{\partial a^*}{\partial x} + \frac{\beta n u}{\delta} \right] dx, \tag{1.73}$$

where $\partial u / \partial x + \partial n / \partial t = 0$. For the soliton they have the values

$$I = \frac{2A^2}{\tilde{k}}, \qquad P = qI + \frac{4}{3} \frac{\beta \delta V A^4}{\tilde{k}(V^2 - c^2)^2} \tag{1.74}$$

The appropriate laws of conservation are written as

$$\begin{aligned} \frac{dI}{dt} &= 2\mathrm{Re} \int_{-\infty}^{+\infty} a[\hat{D}_v(a) + \hat{N}(a)] \, dx, \\ \frac{dP}{dt} &= 2\mathrm{Im} \int_{-\infty}^{+\infty} [\hat{D}_v(a) + \hat{N}(a)] \, dx + \frac{\beta}{\delta} \int_{-\infty}^{+\infty} n \hat{D}_s(u) \, dx. \end{aligned} \tag{1.75}$$

Evolution equations for the parameters A and q may be obtained in the first approximation by substituting the soliton (1.54) into (1.75). Taking into consideration that $u = Vn$ and $n = \delta |a|^2 / (V^2 - c^2)$ for a stationary wave, we find

$$\frac{dI}{dt} = \frac{1}{\pi} \int_{-\infty}^{+\infty} \gamma(k) |a_k|^2 \, dk + 2\mathrm{Re} \int_{-\infty}^{+\infty} a^* \hat{N}(a) \, dx,$$

$$\frac{dP}{dt} = \frac{1}{\pi} \int_{-\infty}^{+\infty} k \gamma(k) |a_k|^2 \, dk + \frac{\beta V}{2\pi \delta} \int_{-\infty}^{+\infty} \Gamma(k) |n_k|^2 \, dk + 2\mathrm{Im} \int_{-\infty}^{+\infty} a^* \frac{\partial}{\partial x} \hat{N}(a) \, dx,$$

where $\quad a_k = \int_{-\infty}^{+\infty} a(x) e^{-ikx} \, dx, \qquad n_k = \int_{-\infty}^{+\infty} n(x) e^{-ikx} \, dx,$

$$\hat{D}_v(a) = \frac{1}{2\pi} \int_{-\infty}^{+\infty} \gamma(k) a_k e^{ikx} \, dk, \qquad \hat{D}_s(n) = \frac{1}{2\pi} \int_{-\infty}^{+\infty} \Gamma(k) n_k e^{ikx} \, dk.$$

For the soliton we have

$$|a_k| = \frac{\pi A}{\tilde{k}\cosh[\pi(k-q)/2\tilde{k}]}, \quad |n_k| = \frac{\pi\delta|k|A^2}{\tilde{k}^2|V^2-c^2|\sinh(\pi|k|/2\tilde{k})}. \quad (1.76)$$

If the spectral dependence of the increments $\gamma(k)$ and $\Gamma(k)$ is slow on the wave number scale $\tilde{k}$, one can make use of the series expansion

$$\gamma(k) = \gamma_0 + \gamma_1(k-q) + \gamma_2(k-q)^2, \qquad \Gamma(k) = \Gamma_0 + \Gamma_1 k + \Gamma_2 k^2 \quad (1.77)$$

and the equations for I and P are simplified substantially. For $\hat{N}(a) = 0$ we have

$$\frac{dI}{dt} = 2I\left(\gamma_0 + \frac{\gamma_2\tilde{k}^2}{3}\right),$$
$$\frac{dP}{dt} = q\frac{dI}{dt} + \frac{4}{3}\gamma_1\tilde{k}A^2 + \frac{2}{3}I\frac{\beta\delta VA^2}{(V^2-c^2)^2}\left(\Gamma_0 + \frac{18\zeta(3)}{\pi^3}\tilde{\Gamma}_1 k + \frac{4}{5}\tilde{\Gamma}_2 k^2\right),$$

where $\zeta(z)$ is the Riemann zeta function. Equations (1.66) for Schrödinger solitons may be obtained from this set under the conditions $\tilde{k} \ll k_w$ and $V \ll c$. In this approximation, system (1.56) may be replaced by Eq. (1.57) if $\Gamma_0 = \Gamma_1 = 0$, with the coefficient being $s = \beta\delta V\Gamma_2/c^4$.

Let us now estimate the effect of the mean flow dissipation described by the coefficient s on the evolution of gravity waves on the surface of a fluid of depth H. If the wavelength λ and the characteristic length of the wave train, L, meet the condition $\lambda \ll H \ll L \sim \lambda/|a|^2$, then the surface waves are described by system (1.56) where $u = \omega n/2k$ is the horizontal velocity component of the long wave, $\beta = \omega/2$, and $\delta = \omega^2/k^2$. The group velocity of deep-water waves, $c_g = \omega/2k \ll c = (gH)^{1/2}$, consequently, the envelope solitons of short waves are described by system (1.56). We will employ the expression $\nu(k) = 2\nu_w k^2$ for viscous damping of surface waves and take as a mechanism governing the wave damping on shallow water the scattering by two-dimensional large-scale bottom inhomogeneities (Pelinovsky, 1979). If the mean-square relative change of the bottom level is $\Sigma = \sqrt{\langle(\Delta H/H)^2\rangle} \ll 1$ and the correlation length of two-dimensional fluctuations is $l \gg L$, then the damping rate is determined by the expression $\Gamma(k) = -c\Sigma^2 l k^2$, so that we have the coefficient $s = -\Sigma^2(l/4k)\sqrt{g/H}$ in (1.56). When $H = 1\,\text{m}$, $\Sigma = 0.1$, and $l = 5\,\text{m}$ at the wavelength $\lambda = 50\,\text{cm}$, the variation of the wave number due to the effect of bottom inhomogeneities on the mean flow equalizes the corresponding change caused by the spectral derivative of the damping rate $d\nu/dk \neq 0$ for the amplitude $A = 0.15$.

The nonlinear evolution of quasi-monochromatic wave packets considered above, including the decay of envelope solitons in which the wave number of a h.f. carrier is adiabaticallydisplaced along the spectrum under the action of weak dissipation or instability, may be manifested in different systems having a nondecaying spectrum. In particular, the evolution of Langmuir solitons interacting with an ion-acoustic wave in collisionless nonisothermal plasma was investigated in the paper (Fabrikant, 1984).

The spectral shift of solitons may be a highly significant element of the mechanism responsible for the onset of self-excited oscillations in a medium with a nondecaying spectrum. Let the instability increment $\gamma(k)$ be positive in a definite spectral interval as the parameter R that describes the deviation of the system from equilibrium (in hydrodynamic flows, in particular, R is Reynolds number) is increasing. The amplitude of the appearing quasi-monochromatic perturbations is small near the instability threshold (in the "soft" regime of saturation). The emerging envelope solitons are displaced towards the maximal spectral dependence of the increment (see Fig. 1.16), eventually forming a periodic modulation wave. As the parameter R increases, the perturbation amplitude grows and it is now necessary to take into account the interaction of the wave packets with the mean flow (low-frequency mode), the dissipation of which gives rise to additional shift of the wave number and, consequently, the solitons escape the region of the maximal increment. Thus, the wave number spectrum now contains a flux of quasi-particles that is determined by the spectral dependence of the wave dissipation and of the mean flow induced by this wave. It should be emphasized that development of such a model for the onset of self-excited oscillations in multidimensional systems encounters severe difficulties caused by the instability of one-dimensional solitons relative to the perturbations distorting the wave front (Yuen and Lake, 1982).

Chapter 2

Wave Generation by Unstable Shear Flows

The theory of hydrodynamic instabilities has been greatly advanced over the past recent decades. The analytical properties and the asymptotic behaviour of perturbations at small viscosity were investigated (Lin, 1955; Drazin and Reid, 1981). Qualitative and numerical methods were developed (Betchov and Criminale, 1967). Considerable progress was recently made in the investigation of nonlinear phenomena (Gollub and Swinney, 1984; Craik, 1985; Rabinovich and Trubetskov, 1989). It is actually impossible to give here even a brief review of all results obtained within this theory. Therefore, we will restrict ourselves only to some problems that are closely allied to the wave propagation in flows.

2.1 Flow stability in ideal fluid. Plasma–hydrodynamic analogy

The amplitude of the oscillations of the stream function ψ for the linear perturbations proportional to $e^{ikx-i\omega t}$ in a two-dimensional shear flow of an ideal incompressible fluid meets the Rayleigh equation

$$\frac{d^2\psi}{dy^2} - k^2\psi - \frac{d^2U/dy^2}{U-\omega/k}\psi = 0, \qquad (2.1)$$

where the $U(y)$ function (the velocity profile) determines an unperturbed flow having the velocity $V_0 = (U(y), 0, 0)$. Equation (1.1) together with boundary conditions must describe the behaviour of small perturbations in a shear flow in the absence of viscosity. The solutions to the corresponding initial-value problem give a set of eigenfunctions and eigenvalues of discrete spectrum that describe the behaviour of small perturbations. There is, however, a significant difficulty associated with the presence in the resonance region of a singular point y_c where the flow velocity is equal to the phase velocity of wave disturbances: $U(y_c) = \omega/k$ (critical layer). In terms of an analytical theory of differential equations (Ince, 1956) the singularity y_c is a logarithmic branching point, provided that y_c does not coincide with the inflection point of velocity profile, i. e., $d^2U(y_c)/dy^2 \neq 0$. Equation (1.1) has an ambiguous general solution; therefore, there arises a problem of the relation between analytical solutions on both sides of y_c.

The description of the dynamics of small perturbations within the Rayleigh equation (1.1) is incorrect in the critical layer; more complicated models must be employed here. That was first done by Lin (1955) who took into account small viscosity and thus removed singularities in the equations of the linear theory. Then, in a viscous fluid, the Rayleigh equation (1.1) transforms into the Orr–Sommerfeld equation (Lin, 1955; Betchov and Criminale, 1967):

$$\frac{i\nu}{k}\left(\frac{d^4\psi}{dy^4} - 2k^2\frac{d^2\psi}{dy^2} + k^4\psi\right) + \left(U - \frac{\omega}{k}\right)\left(\frac{d^2\psi}{dy^2} - k^2\psi\right) - \frac{d^2U}{dy^2}\psi = 0, \tag{2.2}$$

where ν is the kinematic viscosity. The investigation of the asymptotic behaviour of solutions to Eq. (1.2) in the limit $\nu \to 0$, which allows for matching the solutions of the Rayleigh equation (1.1) in the neighbourhood of the singular point y_c, is a problem in the singular perturbation theory (Vazov, 1968). Results of that investigation may be briefly presented as follows.

The solution of the Rayleigh equation that corresponds to growing perturbations is a limiting case of the viscous solution obtained from the Orr–Sommerfeld equation in the limit $\nu \to 0$, given that the branches of a multiple-valued inviscid solution is chosen in the neighbourhood of the branching points y_c using Lin's rule of integration: the integration path around the point y_c in the complex y-plane goes from below if $dU(y_c)/dy > 0$ and from above if $dU(y_c)/dy < 0$.

Integration along a contour in a complex y-plane is a rigorous procedure which gives a unique solution to the problem of hydrodynamic instability of an ideal fluid. This method of solution, however, is purely formal because the variable y is, actually, a real value. There exists an alternative, more physical approach to the choice of the needed branch of solution (Dikiĭ, 1960; Case, 1962). A heuristic role in the development of this approach belongs to the plasma–hydrodynamic analogy, i. e., analysis and comparison of analogous problems and methods in hydrodynamics and plasma physics. Here, we will consider the potentialities of this approach in the interpretation of results obtained in the theory of hydrodynamic instability.

We would like to remind our reader of the results (Galeev and Sagdeev, 1973; Kadomtsev, 1988) known in the theory of amplification (damping) of electromagnetic fields in a plasma. We will restrict ourselves to the simplest case of one-dimensional systems and longitudinal fields.

The values of damping and/or growth of a monochromatic electrostatic field

$$E = (E_x, 0, 0), \quad \text{where} \quad E_x = E_0 \exp(ikx - i\omega t) \tag{2.3}$$

in a system of free electrons with the distribution function $f(r, v) = f_0(v_x)$ may be found by solving a linearized kinetic equation

$$-i(\omega - kv_x)f + \frac{eE_x}{m}\frac{\partial f_0}{\partial v_x} = 0 \tag{2.4}$$

(where e and m are the charge and mass of the electron, respectively), by finding the current $j = e\int f v_x dv_x$, and by determining the conductivity

$$\sigma = \frac{e^2}{m}\int \frac{v_x \partial f_0/\partial v_x}{i(\omega - kv_x)}\, dv_x \tag{2.5}$$

The inequality $\mathrm{Re}\,\sigma < 0$ is the condition necessary for the amplification of the field (1.3). In particular, $\mathrm{Re}\,\sigma$ may be considered to be responsible for the imaginary part of the wave frequency ω (increment): $\gamma = \mathrm{Im}\,\omega \sim -\mathrm{Re}\,\sigma$.

The rule of integration around a singular point in the integral (1.5) (the Landau rule (Landau, 1969)) is obtained by solving the initial-value problem by Laplace transform. The growing ($\mathrm{Im}\,\omega > 0$) field is analysed, and with analytic continuation of the solution to real ω the integration contour in (1.5) must be deformed, go around the singularities in the integrand whilst all the while remaining above them. This gives Landau's rule of integration around singularities for real values of ω and k: the integration contour in (1.5) must go around the singular point $\omega = kv_x$ from below. It can be written as

$$\frac{1}{\omega - kv_x} = \wp\frac{1}{\omega - kv_x} + i\pi\delta(\omega - kv_x), \tag{2.6}$$

where $\wp$ is the principal value sign and $\delta(x)$ is the Dirac delta function.

It should be emphasized that Landau's rule of integration around singular points is thus found in a collisionless approximation, neglecting dissipative effects. Lin's rule in the theory of hydrodynamic instability may also be obtained within an inviscid problem using as a basis the Rayleigh equation (1.1). That was done in the papers (Dikiĭ, 1960; Case, 1962) using a traditional "plasma" method: the initial-value problem was solved by means of Laplace transform. The complex values of ω were considered in the half-plane $\mathrm{Im}\,\omega > 0$, which corresponds to growing perturbations. Lin's rule arises in a natural manner with the analytic continuation of the solution to real values of ω. Note that inclusion of small dissipation (particle collisions) for longitudinal waves in a plasma also removes the resonance singularity in the expression for conductivity, $\sigma(\omega, k)$, and makes it possible to derive Landau's rule of integration around a singular point. Viscous dissipation in hydrodynamics, however, corresponds to the higher-order derivative in the Orr–Sommerfeld equation. An asymptotic solution of a differential equation with a small parameter before higher-order derivative is a rather difficult problem. The Landau method is much simpler and more natural.

Thus, Laplace's method reveals a profound analogy with the plasma problem solved by Landau. This analogy is not restricted to the rule of integration around a singular point. It is also manifested in various wave effects, some of which will be considered below.

The analogy between hydrodynamic and electrodynamic phenomena in flows of particles in the context presented here originated in the work of Case (1962), was then considered by Briggs *et al.* (1970), Timofeev (1970, 1979a,b), Ryutov (1975), and analysed in detail by Andronov and Fabrikant (1979). The analysis showed that the relationship between plasma and hydrodynamic phenomena is more profound than it was noted previously. Below we will show, however, that there are singularities that are specific only to hydrodynamics and do not have simple plasma analogs. One of the most apparent singularities of this type is related to the fact that the velocity at a given point is determined uniquely in classical hydrodynamics so that the particles moving with different velocities must be spaced apart in the medium. Therefore, the simplest wave problems are two-dimensional ones in hydrodynamics. In particular, the particles having different velocities in a shear flow are located at different levels along the y-coordinate which explains the appearance of resonance singular points in the differential equations of the form (1.1).

Let us introduce for a shear flow a quantity that has a sense of the velocity distribution function of particles (Fabrikant, 1976; Mikhaĭlovskiĭ, 1974). Towards this end, we consider in a parallel flow the volume (per unit area in the x, z-plane) occupied by the fluid moving at a velocity ranging from U to $U + dU$:

$$dy = f dU = \frac{dU}{|dU/dy|}. \tag{2.7}$$

It is clear from (1.7) that the quantity $f = |dU/dy|^{-1}$ is the velocity distribution function of particle density in a shear flow. Employing the function $f(U)$ and the plasma–hydrodynamic analogy we can interprete the well-known Rayleigh's theorem (Rayleigh, 1945; Betchov and Criminale, 1967) relating the instability of shear flows in an ideal fluid to the presence of inflection point in the velocity profile. For this purpose we will use the necessary condition for the kinetic instability of longitudinal waves, i. e., the presence of the minimum in the velocity distribution function (Mikhaĭlovskiĭ, 1974). For a parallel flow the extrema of the function $f(U)$, at which $df/dU = 0$ (see Fig. 1.1), are located precisely at the inflection points of the profile $U(y)$, where $d^2U/dy^2 = 0$. Thus the Rayleigh theorem is the hydrodynamic variant of the "plasma" criterion.

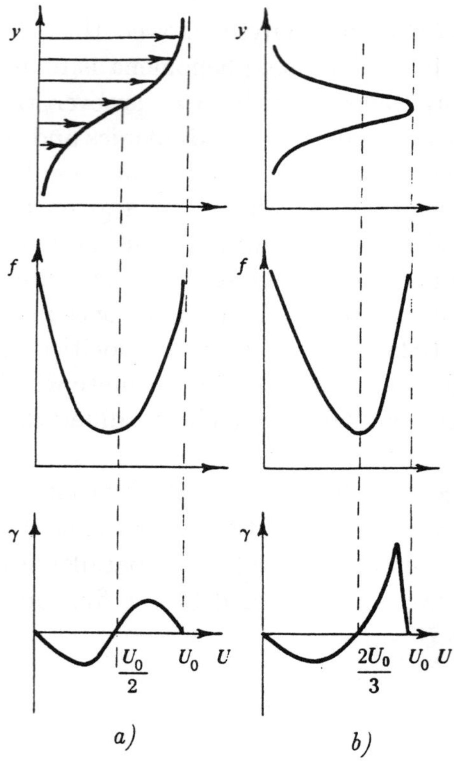

Figure 2.1: Velocity profile $y(U)$ of a shear layer, velocity distribution function $f(U)$ of particles in the flow, and growth rate of the wave interacting with a critical layer: (a) shear layer, (b) jet.

Now consider briefly still another class of the phenomena in shear flows that have a close analogy with the corresponding phenomena in plasma. The singularity of the problem of small perturbations in the absence of dissipation leads to the appearance of singular solutions corresponding to a continuous spectrum. These solutions, that were first found by Van Kampen (1961) in the investigation of the evolution of initial perturbations in a plasma, are freely moving modulated beams of particles. Arbitrary initial perturbation in a plasma may be represented as a superposition of discrete-spectrum waves (Langmuir waves) and continuous-spectrum disturbances damping with time. However, the damping, that is involved with phase scattering of the modulated beams moving with different velocities, occurs by a power law rather than exponentially.

Precisely the same picture was obtained by Case (1962) for the perturbations in a shear flow of an ideal fluid. In this case too, besides the discrete-spectrum solutions that have a definite frequency and increment, there exist solutions meeting the condition $\omega - kU(y) = 0$ where y may take on a continuous spectrum of values. Each of these elementary solutions is a local perturbation of vorticity, i. e., a modulated wave having the vorticity proportional to $\delta[x-U(y)t]$. Arbitrary perturbation in the shear flow decays into a number of discrete-spectrum waves and a continuous spectrum of Van Kampen–Case waves. In a general case, the discrete-spectrum waves do not form a complete system of equations; moreover, there is no discrete spectrum in some flows, for example, in the Couette flow between the planes. The Van Kampen–Case waves move at different phase velocities forming the perturbation damping by a power law.

Note that such a perturbation damping associated with spreading of the beams of particles having different velocities is not related to the increase of entropy and is an inverse process in a system without dissipation. The medium remembers the initial perturbation, the memory being manifested in the effects such as echo that are known in both plasma waves (Kadomtsev, 1988) and perturbations in shear hydrodynamic flows (Nemtsov and Eidman, 1982; Lifshits, 1989).

Thus the damping of continuous-spectrum perturbations determines the evolution of a steady flow; the contribution of the discrete-spectrum waves which decay exponentially is infinitesimal in the limit $t \to \infty$.

At the same time, the shear flow instability depends only on whether Eq. (1.1) has its own solutions belonging to the discrete spectrum so that the frequency lies in the half-plane $\mathrm{Im}\,\omega > 0$. Below, we willconsider a few simple models clarifying in terms of the plasma–hydrodynamic analogy the physical mechanismresponsible for the onset of instability.

2.2 Miles' mechanism for wind wave generation

Consider now the influence of the critical layers on the waves moving in a shear flow. The particles of fluid in the critical layer move synchronously with the wave and may effectively interact with it by either giving up their energy or absorbing the wave energy. Resonance interaction between the waves and the particles in the critical layer may also be interpreted by means of the plasma–hydrodynamic analogy.

We would like to remind our reader that the quantity $\mathrm{Re}\,\sigma$, characterizing the damping of electrostatic field in a plasma, is determined by the half-residue of the integrand in (1.5) that corresponds to the rule of integration around a singular point (1.6):

$$\mathrm{Re}\,\sigma = -\frac{\pi e^2}{m}\left[\frac{\partial f_0}{\partial x}\right]_{v_x=\omega/k} \tag{2.8}$$

Thus the sign of $\mathrm{Re}\,\sigma$ is determined by the derivative of the velocity distribution function, with the particle velocity being equal to the phase velocity of the wave. This is actually connected with the fact that the amount of work performed by a weak wave on the particles is determined by the particles themselves: the ones that overtake the wave or those lagging behind it when $v = c \equiv \omega/k$. If more particles overtake the wave ($\partial f_0/\partial v_x > 0$), then the wave grows. As a result, the waves 1 and 3 (Fig. 1.2) damp and the wave 2 grows in a cold plasma against the background of which a "warm" beam is propagating.

An analogous part is played by the critical layer in a shear flow for wave instability in the systems where the waves are only weakly affected

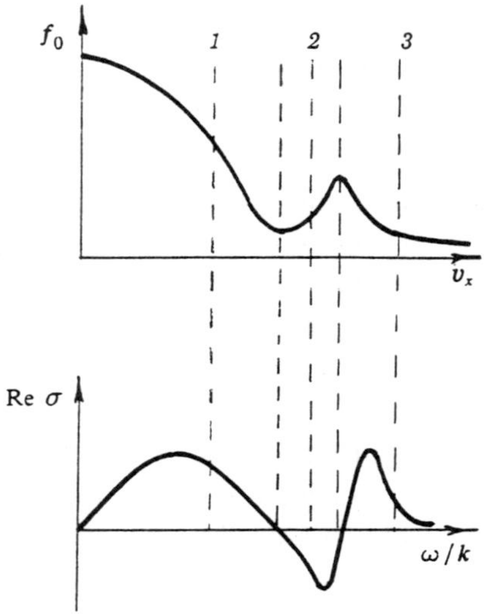

Figure 2.2: Velocity distribution function $f(v)$ of particles and conductivity $\mathrm{Re}\,\sigma$ in a plasma with a beam of electrons.

by the flow. An interesting and important example illustrating the resonance mechanism of hydrodynamic instability in this case is the theory developed by Miles (1957b) explaining the generation of wind waves at a relatively weak wind. Miles' mechanism for wind wave generation has been studied in greatest detail by Barnett and Kenyon (1975) and is in good agreement with experiment in a certain wavelength range (Valenzuela, 1976; Larson and Wright, 1975).

Miles proposed a quasi-laminar model within which an air flow above the water surface is considered as a stationary laminar flow of an ideal fluid. This flow contains small perturbations produced by surface waves on water. Real wind is a turbulent flow but turbulent fluctuations are ignored in the framework of a quasi-laminar model, although the velocity profile of the unperturbed shear flow

$$U(y) = \frac{v_*}{\kappa} \ln\left(\frac{v_* y}{\beta \nu_a}\right), \tag{2.9}$$

is chosen as the profile that is realized for average velocity of a turbulent boundary layer above a smooth solid surface — the so-called logarithmic boundary layer, with $\beta = 0.09$ and $\kappa = 0.4$ are constants, ν_a denotes the kinematic viscosity of the air, and v_* is the characteristic velocity of the boundary layer (Phillips, 1977).

The use of the quasi-laminar approximation leads to a simple model of a shear flow of an ideal fluid above the interface between two media of different densities, the interface "carrying" the surface waves (Fig. 1.3). According to Miles' theory, the density ratio of the air and the water, $\rho_a/\rho_w \simeq 10^{-3}$, is a small parameter which permits use of the perturbation theory. The perturbations of the free surface of a fluid may be considered to the first approximation as wind-independent capillary–gravity waves. In the next-order approximation, one can consider the perturbations introduced by these waves into the air flow, as well as the inverse effect of these perturbations on the water surface oscillations.

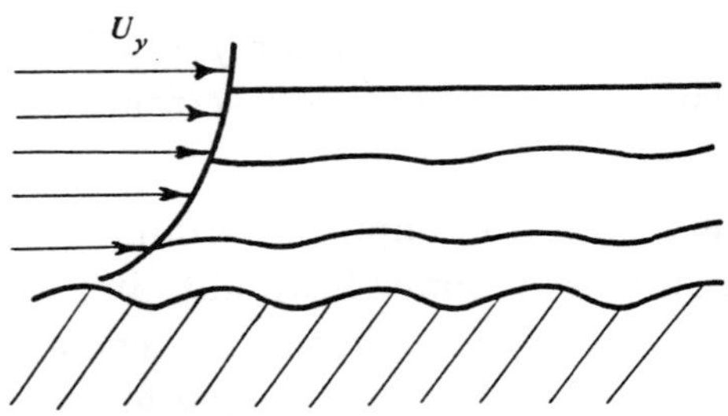

Figure 2.3: Quasilaminar air flow above rough water surface.

The wave perturbations in a shear flow of an ideal fluid are described by the Rayleigh equation (1.1). The boundary conditions for the amplitude of stream function:

$$\psi(+\infty) = 0, \quad \left(\frac{1}{\psi}\frac{d\psi}{dy} + \frac{k}{\omega}\frac{dU}{dy}\right)_{|y=0} = k^2 \frac{Z_w(\omega, k)}{i\omega\rho_a} \tag{2.10}$$

are expressed through the water surface impedance

$$Z_w(\omega, k) \equiv \left(\frac{p}{v}\right)_{|y=+0}.$$

The value of this impedance can easily be found from the boundary condition on the surface of the water:

$$p(-0) = p(+0) + (\rho_w - \rho_a)g\zeta + \sigma\zeta k^2, \tag{2.11}$$

where σ is the coefficient of surface tension. Taking into consideration that the amplitude of potential perturbations in a shallow water approximation decreases with depth as $(\exp(|k|y))$ when $y < 0$, the pressure amplitude in water may be expressed through the amplitude of displacement:

$$p(-0) = \frac{\zeta\omega^2}{|k|}. \tag{2.12}$$

Then we find from (2.11) and (2.12) to the first approximation with respect to the parameter ρ_a/ρ_w the surface impedance of the air-water interface:

$$Z_w = \rho_w \frac{\omega^2/|k| - g - \sigma k^2/\rho_w}{-i\omega} \tag{2.13}$$

Substituting (2.13) into the boundary condition (2.10) for $y = 0$ yields an expression for the instability increment $\gamma \equiv \operatorname{Im}\omega$:

$$\gamma = \frac{-i\omega\rho_a}{k^2(\partial Z_w/\partial\omega)} \operatorname{Im}\left(\frac{1}{\psi}\frac{d\psi}{dy}\right)_{|y=0} = -\frac{\rho_a}{\rho_w}\frac{\omega}{2|k|}\operatorname{Im}\left(\frac{1}{\psi}\frac{d\psi}{dy}\right)_{|y=0}. \tag{2.14}$$

In order to find the quantity $\operatorname{Im}(\psi^{-1}d\psi/dy)_{y=0}$ we first multiply the Rayleigh equation (2.1) by the complex-conjugate function ψ^* and then integrate it in the interval $y > 0$:

$$\left(\psi^*\frac{d\psi}{dy}\right)_{|y=0} = \int_0^\infty \left(\left|\frac{d\psi}{dy}\right|^2 + k^2|\psi|^2 + \frac{d^2U/dy^2}{U - \omega/k}|\psi|^2\right) dy. \tag{2.15}$$

The imaginary part of this relation, with Lin's rule of integration around a singular point in the integrand:

$$\operatorname{Im}\left(\frac{1}{\psi}\frac{d\psi}{dy}\right)_{|y=0} = \pi\left|\frac{\psi(y_c)}{\psi(0)}\right|\left(\frac{d^2U/dy^2}{|dU/dy|}\right)_{|y=y_c}, \tag{2.16}$$

yields

$$\gamma = -\frac{\rho_a}{\rho_w}\frac{\pi\omega}{2|k|}\left|\frac{\psi(y_c)}{\psi(0)}\right|\left(\frac{d^2U/dy^2}{|dU/dy|}\right)_{|y=y_c} \tag{2.17}$$

The sign of the increment is determined by the slope of the velocity profile at the resonance point y_c where $U(y_c) = \omega/k$. In the boundary layer above the water surface, where the velocity profile is convex ($dU/dy < 0$), the resonance interaction with the wind-generated flow amplifies surface waves.

The dependence of the increment on the slope of the velocity profile has a simple physical interpretation. The sign of the second derivative of the velocity profile at the resonance level determines the ratio of the number of particles overtaking the wave to those lagging behind it. If there are more particles overtaking the wave in the neighbourhood of the resonance level, the wave is amplified by taking up the energy from the flow in this neighbourhood. Indeed, the ratio of the number of overtaking and lagging particles depends on the derivative of velocity distribution function of particles in the flow that was introduced in Sect. 2.1:

$$\frac{\partial f}{\partial U} = -\frac{d^2U/dy^2}{|dU/dy|^2} \tag{2.18}$$

The dependence of the wave amplification (damping) on the derivative of velocity distribution function of the particles is analogous to the condition of the plasma wave amplification considered above. This analogy appears to be very natural because both the mechanisms involve the resonance particles moving with a velocity equal to the phase velocity of the wave.

In order to find the value of the increment in an explicit form one needs to know the proper solution of Rayleigh's equation. A relevant numerical computation was performed by Miles (1959). An analytical expression for the increment can be obtained employing the method of matched asymptotic series expansions in a small parameter $\mu = ky_c \ll 1$ (Fabrikant,1980).

It should be emphasized that in most cases important in geophysics, a simple quasi-laminar inviscid Miles model is not complete enough

(Barnett and Kenyon, 1975). As the energy is transferred from wind to waves at strong wind, nonlinear effects (both in the water and in the air) nearly always play a significant part (Hasselmann, 1976). While at strong wind, the air viscosity is essential; in particular, the wind wave increment calculated taking into account air viscosity determines the critical wave number and the critical velocity of the wind (Phillips, 1977). Besides, the turbulence of the wind-generated flow, that is ignored in the quasi-laminar model, may sometimes play a significant role in the generation of wind waves (Phillips, 1977).

2.2.1 Waves on a flowing down sheet

Waves on a flowing down sheet is another example illustrating the Miles mechanism of instability in an ideal fluid. In the reference system related to the unperturbed free surface (Fig. 1.4), the laminar flow of a fluid layer of depth H on a sloping plane has the velocity profile:

$$U = -U_0 \left(\frac{y}{H}\right)^2 \tag{2.19}$$

and is described by the Froude number $Fr = \overline{U}(gh\cos\theta)^{-1/2}$, where $\overline{U} = 2U_0/3$ is the mean flow velocity and θ is the slope of the flow relative to the horizontal plane.

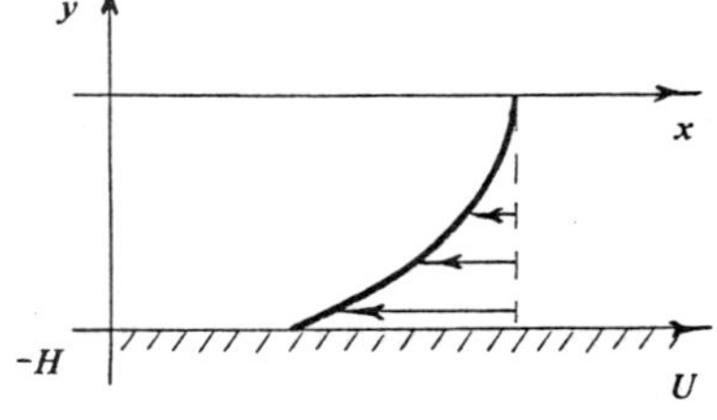

Figure 2.4: Velocity profile of a flowing down sheet.

The wave modes in such a system have proper solutions of the Rayleigh equation (1.1) for $y < 0$ with a parabolic velocity profile and the boundary conditions (1.10) on the free surface $y = 0$ for which

$\rho_a = 0$, $p(+0) = 0$. These proper solutions were found by Petviashvili (1977) who revealed that small perturbations of the free surface are unstable on a flow with a parabolic velocity profile.

It can be shown, however, that the surface waves on a sheet are unstable for an arbitrary convex velocity profile. For this purpose, like it was done above in the problem of wind waves, we will integrate Eq. (1.1) multiplied by ψ^* in the interval $[-H, 0]$, taking into account the boundary conditions on the free surface $y = 0$. As a result we will have

$$\gamma \equiv \mathrm{Im}\,\omega = -\frac{\pi|\omega|^4}{2\mathrm{Re}\,\omega k^2(\rho g + \sigma k^2)} \left|\frac{\psi(y_c)}{\psi(0)}\right|^2 \left[\frac{d^2U/dy^2}{dU/dy}\right]_{y=y_c} \tag{2.20}$$

In the critical layer located below the free layer when $y = y_c < 0$, we have $dU/dy < 0$, $d^2U/dy^2 > 0$. From (1.20) it is clear that $\gamma > 0$, and consequently, the surface wavesgrow.

Note that in allowing for an arbitrary sign of shear flow velocity one should use a general criterion of resonance amplification of waves:

$$U\frac{df}{dU} > 0. \tag{2.21}$$

The amplification is, actually, determined by the energy distribution of particles, i. e., by the quantity $df/d(U^2)$. If the flow velocity has a negative direction, the wave is amplified by the concave portions of the profile that are synchronous to the wave and for which the second derivative $d^2U/dy^2 > 0$.

The complex frequencies for long waves ($kH \ll 1$) were calculated analytically by Petviashvili (1977), where it was shown that they are unstable in a flow having the profile (1.19) for the Froude numbers $\mathrm{Fr} > 0.68$.

Resonance amplification of surface waves interacting with a shear flow in the critical layer does not occur only for wind waves. Miles' theory has been employed to study the generation of acoustic waves on a delay structure (Andronov and Fabrikant, 1979), of magnetohydrodynamic waves in the clouds of radio galaxies (Vrobel' and Kontorovich, 1982), and in jet outbursts in space (Gestrin and Kontorovich, 1986), etc.

Chapter 3

Piecewise-Linear Models of Flows

Different approximations to the velocity profile of the flow which simplify significantly analytical and/or numerical computations are often employed in investigations of hydrodynamical phenomena associated with wave propagation in shear flows. It is most effective to use model flows with piecewise-linear velocity profiles, i. e., the so-called algebraic or Rayleigh's method (Rayleigh, 1945; Betchov and Criminale, 1967).

The simplest flow of this type is a vortex sheet of velocity, when two layers of fluid move without friction so that the tangential velocity component changes in a jump. This classical model of a shear flow has been broadly used for the description of various hydrodynamic phenomena in the atmosphere (Gossard and Hooke, 1975) and in the ocean (Le Blon and Mysek, 1978), in a moving plasma (Mikhaĭlovskiĭ, 1974), in many astrophysical objects (Polyachenko and Friedman, 1976), etc. In spite of its simplicity, it describes quite a number of beautiful and instructive physical phenomena. In particular, the investigation of different mechanisms of hydrodynamic instability within this model may demonstrate the potentialities of the NEW concept presented in Sect. 1.8.

The perturbations growing on the vortex sheet (including the Kelvin–Helmholtz instability in a two-layer fluid of different densities) have received due attention in the literature (e. g., Gerwin, 1968). We will be concerned only with some relatively new and less known results.

3.1 Negative energy waves and Kelvin–Helmholtz instability

The simplest Kelvin–Helmholtz flow is a homogeneous flow of two ideal incompressible fluid layers of different densities in the gravity field with a tangential velocity discontinuity at the layer interface (Fig. 3.1).

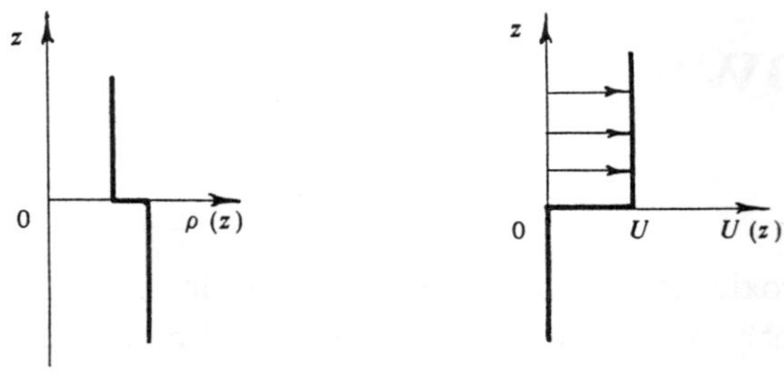

Figure 3.1: Model of the Kelvin–Helmholtz flow in a two-layer liquid with a vortex sheet.

Let, for the sake of definiteness, the lower layer be at rest and the upper layer move with a velocity U. The governing equations of hydrodynamics for linear two-dimensional (only x, z-dependent) perturbations have the following form (Landau and Lifshits, 1986):

$$\frac{\partial u_{1,2}}{\partial t} + U_{1,2}\frac{\partial u_{1,2}}{\partial x} + \frac{1}{\rho_{1,2}}\frac{\partial p_{1,2}}{\partial x} = 0,$$

$$\frac{\partial w_{1,2}}{\partial t} + U_{1,2}\frac{\partial w_{1,2}}{\partial x} + \frac{1}{\rho_{1,2}}\frac{\partial p_{1,2}}{\partial z} + g = 0, \tag{3.1}$$

$$\frac{\partial u_{1,2}}{\partial x} + \frac{\partial w_{1,2}}{\partial z} = 0.$$

Here, u, w are the horizontal and vertical components of velocity vector, respectively, p is the pressure, U denotesthe unperturbed velocity of the flow ($U_1 = U$, $U_2 = 0$); the subscripts 1 and 2 refer to the upper and lower layers, correspondingly. Assuming both the layers to be infinitely deep, we consider all the perturbations to decrease with increasing distance from the layer interface:

$$u_{1,2},\ p_{1,2} \to 0 \quad \text{for} \quad |z| \to \infty. \tag{3.2}$$

The kinematic and dynamic boundary conditions must be satisfied at the interface η (Landau and Lifshits, 1986):

$$\begin{aligned} \frac{\partial \eta}{\partial t} + U_{1,2}\frac{\partial \eta}{\partial x} &= w_{1,2}, \\ p_1 - p_2 &= (\rho_2 - \rho_1)g\eta - \sigma\frac{\partial^2 \eta}{\partial x^2}. \end{aligned} \tag{3.3}$$

In a linear approximation, however, it suffices to meet these conditions at the unperturbed boundary $z = 0$.

Writing down the solution of the system (3.1) in the form

$$\left\{ \begin{array}{c} u_{1,2} \\ w_{1,2} \\ p + \rho_{1,2}gz \\ \eta \end{array} \right\} = \left\{ \begin{array}{c} A_{1,2}e^{kz} \\ -iA_{1,2}e^{kz} \\ \rho_{1,2}(\omega/k - U_{1,2})A_{1,2}e^{kz} \\ \eta_0 \end{array} \right\} e^{i(kx-\omega t)} \tag{3.4}$$

and substituting it into (3.1) one obtains the following dispersion equation (Landau and Lifshits, 1986):

$$\omega = \frac{a}{1+a}kU \pm \left[\frac{1-a}{1+a}gk + \frac{\sigma k^2/\rho_2}{1+a} - \frac{a}{(1+a)^2}k^2U^2\right]^{1/2}. \tag{3.5}$$

Here, $a = \rho_1/\rho_2$ may be an arbitrary value that is restricted only by the condition of static stability $0 \le a \le 1$, σ is the coefficient of surface tension,and g is the acceleration due to gravity.

For $U = 0$, Eq. (3.5) describes two branches of dispersion curves that are symmetric with respect to the k-axis and correspond to standard capillary-gravity waves travelling in opposite directions (Fig. 3.2, curve

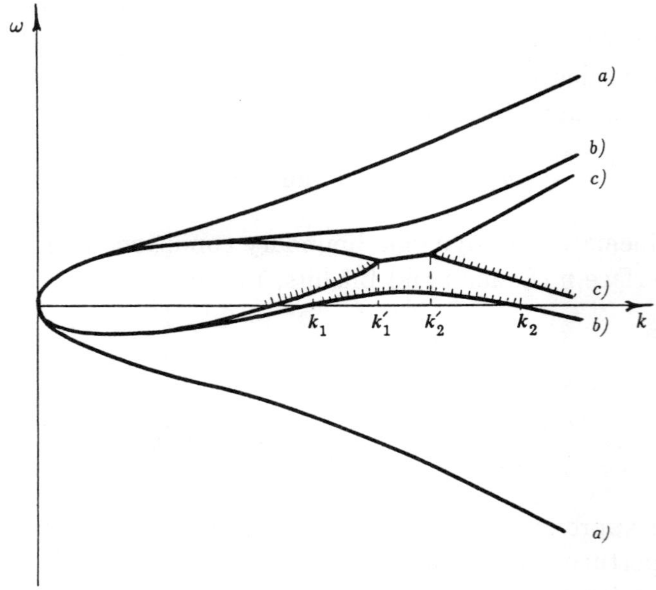

Figure 3.2: Dispersion curves for the Kelvin–Helmholtz model:
(a) – $U = 0$, (b) – $U_{c1} < U_{KH}$, (c) – $U > U_{KH}$. Hatched are the portions corresponding to NEWs.

a). [1] Whereas for $U \neq 0$, the dispersion curves become nonsymmetric owing to the wave drift caused by the flow. For

$$U > U_{c1} \equiv \left[4 \frac{1-a}{a^2} \frac{\sigma g}{\rho_2} \right]^{1/4}, \tag{3.6}$$

the lower branch of the dispersion curve (Fig. 3.2) changes its sign and becomes positive in the portion $k_1' \leq k \leq k_2'$, where

$$k_{1,2}' = \frac{\rho_1 U^2}{2\sigma} \left[1 \pm \sqrt{1 - \frac{U_{c1}^4}{U^4}} \right].$$

[1]We assume for convenience that the wave number k of the perturbations is positive, while the frequency ω may have an arbitrary sign. Alternatively, we could assume $\omega \geq 0$ and $k \in (-\infty, \infty)$, but further considerations will then be less illustrative and more complicated.

According to what was said in Sect. 1.8, this portion of the curve corresponds to negative energy waves. As U is increased further, the two branches continue to converge; finally, when

$$U \geq U_{KH} \equiv U_{c1}\sqrt{1+a}, \tag{3.7}$$

they reconnect (see Fig. 3.2, branch c) and the known Kelvin–Helmholtz (KH) instability arises in the region $k_1 \leq k \leq k_2$, where

$$k'_{1,2} = \frac{\rho_1 U^2}{2\sigma(1+a)}\left[1 \pm \sqrt{1-(1+a)^2\frac{U_c'^4}{U^4}}\,\right]. \tag{3.8}$$

This reconnection is typical of the interaction of the waves having different energy signs (see Sect. 1.8). Thus, the appearance of the KH instability can be attributed to the interaction of the waves whose energies have different signs: the NEWs corresponding to the lower branch of the dispersion curve give up their energy to the positive energy waves corresponding to the upper branch, with the amplitudes of the waves of both signs growing.

It is interesting that, even in the presence of the KH instability, the system under consideration contains nongrowing but potentially unstable NEWs that lie in the intervals between the points where $\omega_2 = 0$ and $d\omega_2/dk = \infty$. The waves will grow as their energy is depleted. One can easily see from the expressions for U_{c1} and U_{KH} that these velocities are almost equal if a is small, as is the case of the water–air interface, where $a \simeq 10^{-3}$. However, if $a \simeq 1$, which is typical, for example, for internal layers in the ocean or atmosphere, then $U_{c1} \simeq U_{KH}/\sqrt{2}$ (Benjamin (1963) was the first to note this circumstance). There is usually no tension ($\sigma = 0$) in internal layers, therefore, within the ideal model considered here the KH instability occurs at arbitrary values of U for the waves with $k > k_1 = (1-a^2)g/aU^2$ (in the real ocean, the instability is suppressed at large k due to viscosity and finiteness of the thickness of the shear layer), while the NEW region is located at smaller values of k: $k'_1 = k_1/(1+a) < k < k_1$, i. e., in the region of longer waves.

3.2 Dissipative instability of a vortex sheet

There exist different mechanisms responsible for the growth of NEWs outside the region of KH instability by depleting their energy. Thus, analogously to the plasma dissipative instability (Nezlin, 1976), viscous dissipation in the lower layer at rest leads to growth of NEWs (Cairns, 1979).

It is important that for the NEWs to be amplified the viscosity must lead to "positive losses". For example, the NEWs in the model considered are damped if the upper (instead of the lower) moving layer is viscous. In this case, however, positive energy waves may grow on the upper branch of the dispersion curve, since the viscosity in the moving upper layer leads to "negative damping". Indeed, in passing over to the reference system where the upper layer is at rest, the energy of the growing mode and, simultaneously, the dissipation change their signs (see Sect. 1.8). In this reference system, NEWs on the upper branch of the dispersion curve now grow under the action of positive dissipation, the presence of instability at this mode, evidently, being invariant to the choice of reference system.

It should be emphasized that dissipative instability occurs on both branches of the dispersion curve for the waves that are retarded relative to the flow, their phase velocity being smaller than the phase velocity of the corresponding fluid layer. In a general case, the critical value of the relative velocity of the layers, at which the instability may occur on the upper branch, can be written as

$$U_{c2} = \sqrt[4]{(1-a)g\sigma/\rho_2}. \tag{3.9}$$

This value differs from U_{c1}. The difference between the critical velocities for the upper (light) and lower (heavy) fluids is small for $a \simeq 1$ (internal waves in the ocean pycnocline), but it is rather significant for $a \ll 1$ (surface water waves).

Let us demonstrate what was said above graphically, passing over to the "mean-mass" reference system uniformly moving with the velocity

$V = aU/(1+a)$. In this reference system, the lighter fluid moves in the positive direction with velocity $U/(1+a)$, while the heavier fluid moves in the opposite direction with a smaller velocity $aU/(1+a)$. The branches of the dispersion curve here are symmetric relative to the k-axis due to the Dopler frequency shift $\Omega = \omega - akU/(1+a)$ (Fig. 3.3). Note that there are no NEWs in this reference system (the dispersion curves do not change their sign and $c_g = d\omega/dk \neq \infty$ for $U < U_{KH}$).

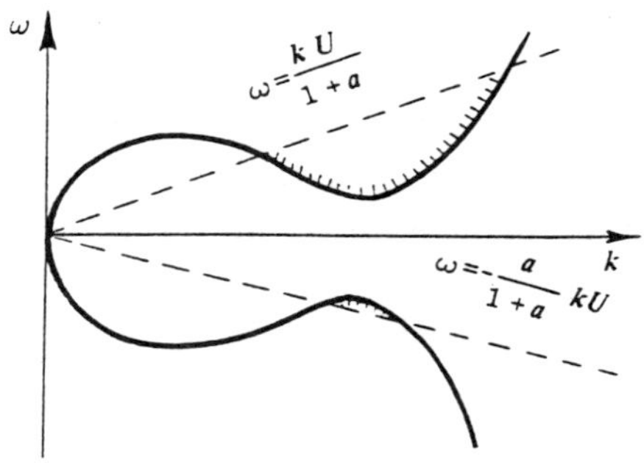

Figure 3.3: Dispersion curves for waves on a vortex sheet in a "mean mass" reference system. The hatched portions correspond to the "potentially" unstable waves for which dissipative instability is possible.

For sufficiently large values of U, there appear on the dispersion curve portions with retarded waves; in Fig. 3.3 they are marked off by rays with the slope $U/(1+a)$ (in the upper half-plane) and with the slope $-aU/(1+a)$ (in the lower half-plane). Apparently, as the velocity U of the relative motion of layers increases, the retarded waves first appear on the upper branch of the dispersion curve.

Consider a particular example. The estimates made for the water-air interface give $U_{KH} \approx U_{c1} \approx 6.5\,\mathrm{m/s}$ and $U_{c2} \approx 0.23\,\mathrm{m/s}$ which is equal to the minimal phase velocity of surface waves on water. For $U > U_{c2}$, the air viscosity ensures negative dissipation, which may give rise to the dissipative instability of surface waves. It should be taken into consideration, however, that the dissipation in a heavy medium (water) persists to be positive up to the velocities $U > U_{c2} \approx U_{KH}$. Thus, the dissipation in water increases the threshold velocity of the wind and, consequently, delays the onset of instability of wind waves (Phillips, 1977).

As the velocity is further increased, the dissipation in the lower layer also changes its sign for $U > U_{c2}$ and then, almost immediately, for $U > U_{KH}$ ($\approx U_{c2}$ if $a \ll 1$) the dispersion branches reconnect and the KH instability arises. In real conditions, wind wave instability occurs much earlier, at $U \approx 1.3\,\mathrm{m}/s \ll U_{KH} \approx 6.5\,\mathrm{m/s}$.

It follows from what was said above that for $\rho_1 \ll \rho_2$ the lower layer of heavy fluid "carries" the surface waves whose dispersion properties (for $U \ll U_{KH}$) are perturbed only slightly under the action of the upper layer of the light medium. At the same time, the dissipation in the air that is moving faster than the wave and is changing its sign gives rise to the development of instability of surface waves, i. e., to the appearance of wind waves.

Let us evaluate the dissipative instability of surface waves within the vortex sheet model assuming the fluid in the lower layer to be ideal but taking into account viscosity in the upper layer (air). This approach is justified if the dynamic air viscosity (that is considered to be turbulent) is much higher than the dynamic viscosity of water. A dispersion equation for the waves travelling at an angle θ to the direction of the wind may be readily derived, taking into account the equality of tangential and normal stresses at the interface between media:

$$a(\omega - \mathbf{kU})^2 + \omega^2 - (1-a)gk - \sigma k^3/\rho_2 = -4ik^2 a\nu_t(\omega - \mathbf{kU}), \quad (3.10)$$

where ν_t is the coefficient of turbulent kinematic viscosity of the air $k = |\mathbf{k}|$. Assuming the imaginary correction to frequency, $\mathrm{Im}\,\omega$, due to viscosity to be small compared to the real frequency $\mathrm{Re}\,\omega$, we obtain from (3.10) an expression for the growth rate of waves:

$$\mathrm{Im}\,\omega = \frac{-2a\nu_t k^2}{1+a}\left[1 - \frac{\mathbf{kU}}{\sqrt{(1-a^2)gk + \sigma(1+a)k^3/\rho_2 - a(\mathbf{kU})^2}}\right]. \quad (3.11)$$

Thus, in this reference system, the dissipation is negative and causes amplification of the surface waves having positive energy.

It should be emphasized that the change in the sign of the dissipation for $\omega - kU < 0$ is not related to its specific physical mechanism. In particular, the wind instability of surface waves could be due to different factors, including the fact that viscosity (possibly, turbulent) in the wind-generated flow may play a significant role (Barnett and Kenyon, 1975), together with Miles' mechanism of instability and other well-known mechanisms.

Analysis of the formula (3.11) shows that the waves travelling along the flow grow in the wave number interval

$$\Delta k = k_2'' - k_1'' = \frac{\rho_2 U^2}{\sigma}\sqrt{1 - \frac{4g\sigma(1-a)}{U^4\rho_2}}, \tag{3.12}$$

where

$$k_{1,2}'' = \frac{\rho_2 U^2}{2\sigma}\left[1 \pm \sqrt{1 - \frac{4g\sigma(1-a)}{U^4\rho_2}}\right]. \tag{3.13}$$

The waves travelling at an angle to the flow are amplified only in a narrow angular region whose boundaries are defined by the equality

$$\cos\theta_{\max} = \frac{\sqrt[4]{4(1-a)g\sigma/\rho_2}}{U} \equiv \frac{U_{c2}}{U}. \tag{3.14}$$

Outside this interval the waves are damped.

The dependence $\theta_{\max}(U)$ is plotted in Fig. 3.4. For comparison, Fig. 3.4 also shows the dependence $\theta_{\max}(U)$ (at two values of the parameter a) for the KH instability that can be found from the formula (Stepanyants, 1983):

$$\cos\theta_{\max} = \frac{U_{c2}\sqrt{1+1/a}}{U} \equiv \frac{U_{KH}}{U}. \tag{3.15}$$

It can be said, in conclusion, that the instability of retarded waves (or NEWs) arises at much smaller flow velocities and leads to wave amplification in a wide angular region when the KH instability has not yet appeared. This permits one to distinguish between these two instabilities experimentally.

Unfortunately, there are very few known hydrodynamic experiments with NEWs, unlike plasma physics where dissipative instabilities have

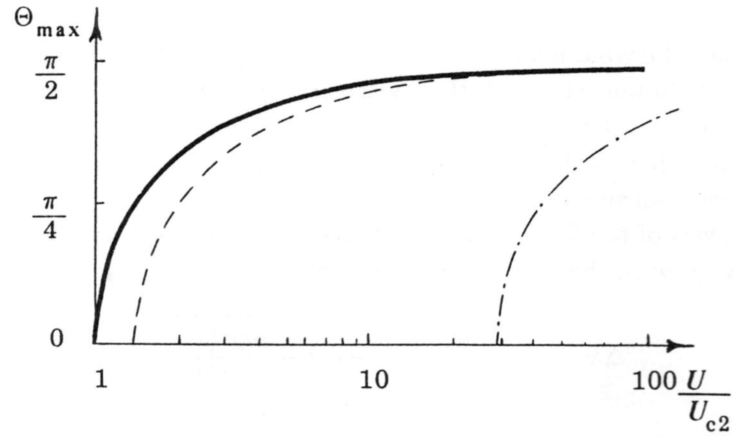

Figure 3.4: The angle boundaries of the region of wind wave amplification versus wind velocity in the presence of dissipative instability (solid curve) and Kelvin–Helmholtz instability (dashed curve for $a \simeq 1$ and dot-dash curve for $a = 0.0012$).

long been known (Mikhaĭlovskiĭ, 1974). Leaving aside the numerous works on the excitation of the Tollmien–Schlichting waves in boundary layers (near water surface including), which according to Benjamin (1963) also have negative energy, we will call your attention only to the NEWs observations in free layer flows.

An interesting laboratory experiment was carried out by Komatsu *et al.* (1978). The authors investigated the KH instability at the interface between fresh and saline (heavier) water in a tank. They really observed KH instability in the portion of the dispersion curve where it is expected to be according to the theory.

Besides, they discovered in those experiments a different type of instability that can be interpreted as NEW instability. Indeed, let us consider Fig. 3.5 from their paper where the frequency dependence of the phase velocity of waves is depicted. This is, actually, a graphical representation of the dispersion relation (3.5) without surface tension ($\sigma = 0$). The points designate the parameters of the growing perturbations observed in the experiment. Portion 2 of the curve corresponds to the KH instability.

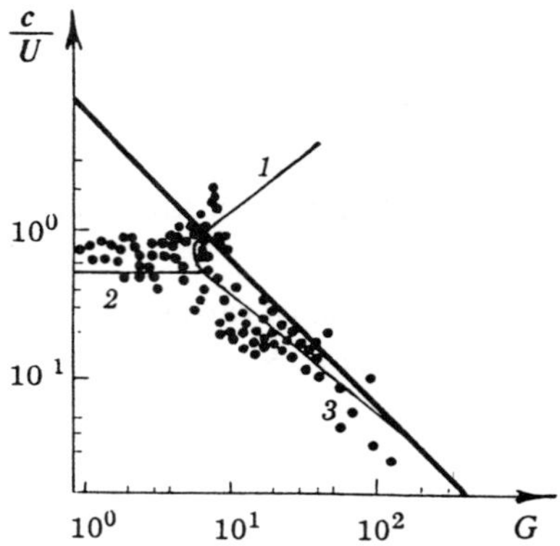

Figure 3.5: Theoretical dispersion curves (normalized phase speed c/U versus parameter $G \equiv 2\pi(1-a)g/\omega U$) for waves on a vortex sheet in a two-layer liquid and experimental data (dots) corresponding to amplified waves (courtesy of Komatsu *et al.*, 1978).

One can also see other growing perturbations whose frequencies differ by more than an order of magnitude from the frequency range of KH instability. It can readily be shown that these perturbations are retarded and have negative energy in the appropriate reference system.

The boundary of the region of retarded waves is determined from (3.5) and can be written as (Stepanyants, 1983)

$$\frac{c}{U} < \frac{\omega U}{g(1-a)}, \tag{3.16}$$

where $c = \omega/k$. The region below the heavy curve in Fig. 3.5 corresponds to the inequality (3.16). Clearly, the parameters of the observed amplified waves correspond to the retarded waves (portion 3 of the curve) which, presumably, grow due to dissipative instability.

3.3 Radiative instability

The mechanism responsible for the energy sink of NEWs on a vortex sheet between two fluid layers having different densities deserves separate consideration. This mechanism involves emission of bulk internal or acoustic waves on a vortex sheet. The arising instability is referred to as radiative instability. As far as we know, this type of instability was first considered in (Erokhin and Moiseev, 1974) as applied to plasma waves. Later, it was also studied in hydrodynamics and acoustics (Ostrovsky *et al.*, 1986).

Here we explain the mechanism of radiative instability on an example a slightly modified model of a vortex sheet (Fig. 3.6a).

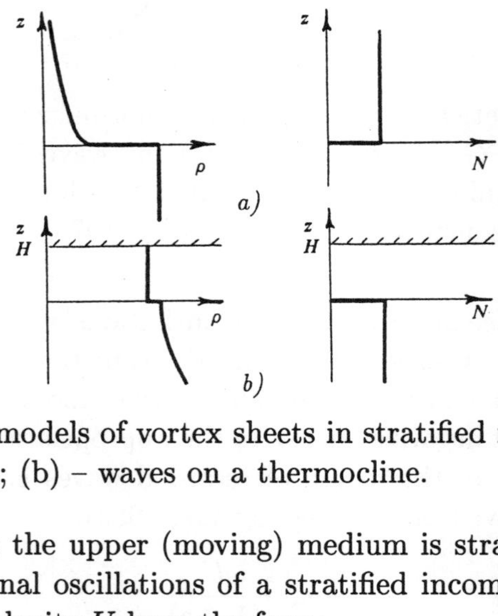

Figure 3.6: The models of vortex sheets in stratified media: (a) – wind waves; (b) – waves on a thermocline.

Suppose that the upper (moving) medium is stratified. Equations for two-dimensional oscillations of a stratified incompressible medium moving with a velocity U have the form:

$$
\begin{aligned}
\frac{\partial u}{\partial t} + U\frac{\partial u}{\partial x} + \frac{1}{\rho}\frac{\partial p}{\partial x} &= 0, \\
\frac{\partial v}{\partial t} + U\frac{\partial U}{\partial x} + \frac{1}{\rho}\frac{\partial p}{\partial z} + \frac{g\rho}{\rho} &= 0, \\
\frac{\partial \tilde{\rho}}{\partial t} + U\frac{\partial \tilde{\rho}}{\partial x} + v\frac{\partial \rho}{\partial z} &= 0, \\
\frac{\partial u}{\partial x} + \frac{\partial v}{\partial z} &= 0,
\end{aligned}
\tag{3.17}
$$

where ρ is unperturbed density and $\tilde{\rho}$ is its perturbation.

Let us employ the Boussinesq approximation that corresponds to the limiting transition $d\rho/dz \to 0$, $g \to \infty$, but so that the Brunt–Väisälä frequency $N \equiv [(-g/\rho)d\rho/dz]^{1/2} = \text{const}$. The density changes in this case insignificantly on the wavelength scale, but the restoring buoyancy force is retained. In this approximation, the derivative $d\rho/dz$ enters only in N and ρ can be assumed to be constant in the coefficients of the equations. The equations for small two-dimensional perturbations proportional to $\exp(-i\omega t + ikx + iqz)$ in the upper fluid layer moving with velocity $\mathbf{U} = (U, 0, 0) = \text{const}$ in the Boussinesq approximation have the form

$$\begin{aligned} -i(\omega - kU)u + ikp &= 0, \\ -i(\omega - kU)v\left[1 - \frac{N^2}{(\omega - kU)^2}\right] + \frac{\partial p}{\partial z} &= 0, \\ iku + \frac{\partial v}{\partial y} &= 0. \end{aligned} \tag{3.18}$$

One can now derive dispersion equations for plane waves and their group velocity $\mathbf{c}_g$:

$$\begin{gathered} (\omega - kU)^2 = \frac{N^2k^2}{k^2 + q^2} = N^2 \sin^2\theta, \quad \mathbf{k} = k_0(\sin\theta, \cos\theta), \\ \mathbf{c}_g = \frac{Nq}{(k^2+q^2)^{3/2}}(q_1, -k) = \frac{N\cos\theta}{k_0}(\cos\theta, -\sin\theta), \\ (\mathbf{k}\cdot\mathbf{c}_g) = 0, \quad q(\omega, k) = k\sqrt{N^2/\omega^2 - 1}. \end{gathered} \tag{3.19}$$

Thus, the presence of stratification determines a continuous spectrum of the wave perturbations due to the buoyancy force acting on the fluid particles of different densities.

For $\omega > N$, the amplitude of perturbations changes exponentially along the z-coordinate because, in this case, $q(\omega, k)$ is an imaginary value. The boundary condition in the limit $z \to \infty$ reduces to the inequality $\operatorname{Im} q(\omega, k) > 0$.

At the frequencies $\omega < N$, the perturbations are propagating IGWs. The condition of radiation — the group velocity of the escaping wave must propagate from the vortex sheet — is in this case the boundary condition, as $z \to \infty$. This condition allows for the sign of the vertical component of the wave vector in the moving medium to be chosen as $q(\omega, k) < 0$.

Matching, like before, the amplitudes of vortex sheetperturbations at $z = 0$ yields the following dispersion equation for waves on discontinuity:

$$a(\omega - kU)^2\kappa_1 + \omega^2 = (1 - a)gk + \sigma k^3/\rho_2, \tag{3.20}$$

where $\kappa_1 = iq_1/k = \sqrt{1 - N^2/(\omega - kU)^2}$.

Consider as a possible application of such a model the effect of atmospheric stratification on wind waves. Assuming $a \ll 1$, one finds from Eq. (3.20) to a first approximation in the parameter a a correction to the capillary-gravity waves. This correction has the imaginary part for $|\omega - kU| < N$:

$$\operatorname{Im}\omega = \frac{a(\omega - kU)}{2\omega}\sqrt{N^2 - (\omega - kU)^2}. \tag{3.21}$$

The value of $\operatorname{Im}\omega$ is negative for sufficiently small wind velocities ($U < U_{c1}$), which corresponds to surface wave damping. The mechanism of damping is associated with emission of internal gravity waves (IGW). These leaky waves carry off the energy of the capillary-gravity wave, which leads to its damping. The effect of surface wave damping caused by the emission of bulk leaky waves is well-known for waves of different types (Goldstone and Oliner, 1959).

Note that, for $\operatorname{Im}\omega < 0$, the vertical wave-vector component of the IGW moving from the density discontinuity has a negative imaginary part $\operatorname{Im} q < 0$. Consequently, the amplitude of the wave field grows exponentially (rather than decreases) upwards from the $z = 0$ surface. The nonlocalized solution obtained in this fashion is not quite correct. But the physics of the field increase far from density discontinuity is clear from a nonstationary (initial-value) problem. The later the surface wave damping with time emits a leaky wave escaping from the discontinuity, the smaller is its amplitude. As a result, the amplitude of the wave field far from the surface of discontinuity is larger than near it.

The surface wave may grow ($\mathrm{Im}\,\omega > 0$) at sufficiently large velocities, $U > U_{c1}$. This instability exists only in a narrow range of frequencies:

$$kU - N < \omega < kU. \tag{3.22}$$

The instability mechanism consists in emission of the leaky waves possessing negative energy. Note that the threshold velocity of radiative instability of wind waves, similarly to the mechanism of negative viscous dissipation considered above, is, like before, equal to $U_{c1} \approx 23$ cm/s. Therefore, even small stratification in the upper layer may be sufficient for the growth of wind waves. The stratification in the lower layer (water) may also initiate radiative instability (Benjamin, 1963). But this is possible only at much larger wind velocities, $U > U_{c2}$, when NEWs appear on the lower branch of the dispersion curve (see Sect. 3.1).

Consider another example demonstrating the onset of radiative instability that may have practical use. A vortex sheet model with density discontinuity may be employed for qualitative description of a real structure of the upper ocean. We suppose that a vortex sheet with density discontinuity (pycnocline) separates the moving layer $0 < z < H$ of a homogeneous fluid of finite thickness H (the upper mixed layer in the ocean) and the unbounded stratified medium ($z < 0$) having an unperturbed velocity $U = 0$ and a Brunt–Väisälä frequency $N = \mathrm{const}$ (Fig. 3.6b). The boundary condition $w = 0$ is fulfilled on the $z = H$ surface. This condition is referred to as the rigid top approximation and is often used in the description of the dynamics of internal waves in the ocean.

The dispersion equation for surface waves on a vortex sheet in this case has a form (Ostrovsky *et al.*, 1983)

$$a(\omega - kU)^2 \coth(kH) + \omega\sqrt{\omega^2 - N^2} - (1 - a)gk. \tag{3.23}$$

This equation must be supplemented with a boundary condition in the limit $z \to -\infty$ which, like in the problem considered above for wind waves, reduces either to the amplitude decrease with increasing distance from the vortex sheet (if $\omega > N$) or to the radiation condition for IGW (if $\omega < N$). This boundary condition determines the choice of the branch of the multivalued function in the dispersion equation (3.23).

The dispersion curve corresponding to the solution of (3.23) is presented in Fig. 3.7 for sufficiently large velocities

$$U > \tilde{U}_c \equiv \sqrt{\frac{1-a}{a} gH}.$$

The KH instability here occurs in the region of rather short waves and in the real ocean it may be restricted, for instance, by a finite thickness of the shear layer. At the same time, the radiative instability covers a longer wavelength range: $0 < k < k_N$. Two instability regions here are separated by a "transparency window" inside which $\mathrm{Im}\,\omega = 0$ and perturbations do not grow.

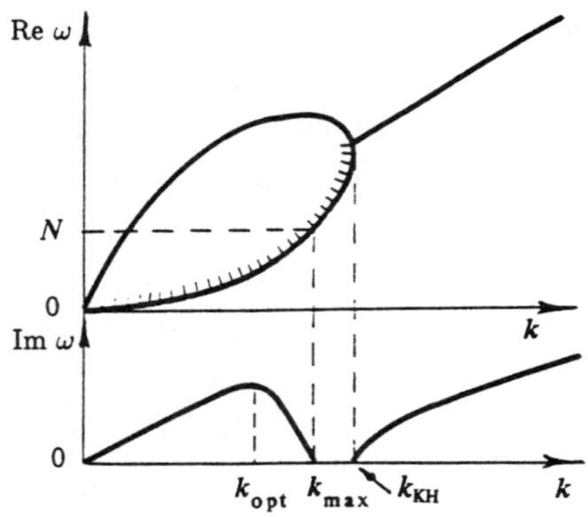

Figure 3.7: Dispersion curves for the waves on a vortex sheet in a two-layer liquid in the presence of radiative instability.

The long waves on the surface of the vortex sheet have negative energy in this case and meet the condition $\omega - kU < 0$. The instability occurs as the leaky waves (IGWs with positive energy) are emitted deep into the ocean, which ensures the mechanism of effective positive dissipation for surface NEWs.

Besides, the dissipation in a moving layer gives rise to wave damping. The mixed layer in the ocean is usually strongly turbulent, therefore, the model must take into account, in the first place, turbulent dissipation in the upper ocean. Ostrovsky and Soustova (1979) showed that internal wave damping in the ocean is governed by two mechanisms. One of them is involed with turbulent viscosity and the other, with the expenditure of internal wave energy on sustaining the turbulence in the field of buoyancy force. The total damping rate found from a linearized system of semiempirical equations for averaged motions and energy of turbulent pulsations has the form

$$\Gamma(k) = -\frac{\Pi}{gl_z(1-a)} - \nu_t k^2, \tag{3.24}$$

where Π is turbulent buoyancy flow, l stands for the vertical scale of turbulence, and ν_t for the coefficient of turbulent viscosity.

The total increment of long surface waves on thermocline, that is the difference between the increment of radiative instability and the rate of turbulent damping, may be approximated by a cubic parabola:

$$\gamma(k) = \gamma_0 + \gamma_1|k| + \gamma_2 k^2 + \gamma_3|k|^3. \tag{3.25}$$

This approximation gives a good description of the qualitative dependence of the increment on the wave number and differs slightly from a more exact $\gamma(k)$ dependence that can be obtained from a numerical solution of the dispersion equation for waves on a pycnocline (Ostrovsky *et al.*, 1983).

3.4 Nonlinear dynamics of long waves in hydrodynamic flows

The mechanisms of instability in flows with a vortex sheet that were considered above hold in a broad spectral interval starting from the long waves for which $k \to 0$. Note that the low-frequency perturbations with the wavelength greater than the characteristic size of the layer are, to a certain extent, "rough" in that their properties are determined only by integral parameters of the flow and weakly depend on details of velocity and density profiles. Nonlinear evolution of growing long-wave perturbations may also be investigated in a general form (Rabinovich and Fabrikant, 1976). Unlike the classical Kelvin–Helmholtz instability, a weaker dissipative instability of long waves does not necessarily give rise to breaking and turbulent excitation of wave crests; the growth of the wave amplitude may be restricted by nonlinear dissipation at a low level, thus resulting in the formation of stationary travelling waves.

Weakly nonlinear flow perturbations can be described in many cases of practical interest in the so-called single-wave approximation, when, by virtue of small nonlinearity, the consideration may be restricted to wave perturbations of one type (e. g., a nonlinear wave travelling in one direction), ignoring the interaction of the waves that belong to different branches of the dispersion curve. Formal description of travelling waves will then reduce to analysis of solutions to a partial differential equation that is, as a rule, a generalization of one of the well-known reference equations (Burgers, Korteweg–de Vries, Benjamin–Ono, or other equations) to nonequilibrium media.

One of the most general approximate equations describing the evolution of long-wave perturbations in a medium with quadratic nonlinearity has a form

$$\frac{\partial \eta}{\partial t} + c\frac{\partial \eta}{\partial x} + \alpha\eta\frac{\partial \eta}{\partial x} - \beta\frac{\partial^3 \eta}{\partial x^3} = \gamma_0\eta + \frac{\gamma_1}{\pi}\frac{\partial}{\partial x}\int_{-\infty}^{\infty}\frac{\eta(x',t)\,dx'}{x-x'}$$

$$+ \gamma_2\frac{\partial^2 \eta}{\partial x^2} + \frac{\gamma_3}{\pi}\frac{\partial^3}{\partial x^3}\int_{-\infty}^{\infty}\frac{\eta(x',t)\,dx'}{x-x'}. \tag{3.26}$$

Here, η is the deviation of the interface between two fluids and the integrals on the left-hand side are understood as principal value integrals.

Equation (3.26) can be derived from general equations of hydrodynamics using one of the variants of the method of multiscale series expansions in small parameters of nonlinearity, dispersion and dissipation (Ostrovsky, 1974; Pelinovsky and Friedman, 1974). However, taking into consideration that the effects of nonlinearity, dispersion and dissipation to the first approximation enter the evolution equation additively, one can find first the nonlinear term, and the linear terms associated with dispersion and dissipation may be readily obtained proceeding from the series expansion of complex frequency in the wave number. Then, all the linear terms in Eq. (3.26) are formally obtained by substituting $i\omega \to -\partial/\partial t$ and $ik \to \partial/\partial x$ into the dispersion equation (Kornel and Banergy, 1984). Besides, the terms proportional to k and k^3 in the series expansion of the increment (3.25) are described by the Gilbert transform of the first and third derivatives of the functions $\eta(x,t)$ (Tsimring, 1982).

Two principal types of spectral dependences of the increment, $\gamma(k)$, may be distinguished. The first of them corresponds to long-wave instability, when the maximum of the increment falls at the wave number $k = 0$ (Fig. 3.8a). The onset of waves is connected here with the energy

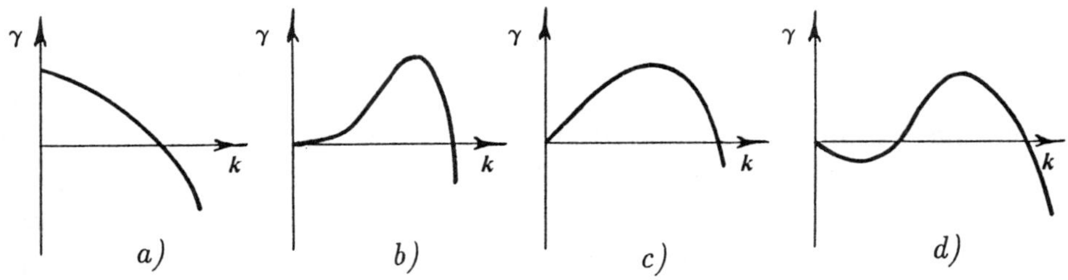

Figure 3.8: Different types of dependences of the growth rate of long-wave instability on wavenumber.

transfer (due to nonlinearity) from the instability region upwards of the wave number spectrum and with subsequent dissipation of higher harmonics. The other instability type is the so-called "negative viscosity", i. e., the growth of the increment with the increase of the wave number. Below, we will consider some solutions to Eq. (3.26) that describe the dynamics of nonlinear waves for different $\gamma(k)$ dependences.

3.4.1 The evolution of waves in media with long-wave instability

Consider first the simplest example illustrating nonlinear effects in unstable media without dispersion. Let us take a particular case of equation (3.26):

$$\frac{\partial \eta}{\partial t} + \alpha\eta\frac{\partial \eta}{\partial x} = \gamma_0\eta + \gamma_2\frac{\partial^2 \eta}{\partial x^2}. \tag{3.27}$$

This equation describes, in particular, capillary instability of liquid jets at small Reynolds numbers (Novikov, 1977); it can be used to follow the combined effect of instability, nonlinearity, and high-frequency dissipation on the evolution of a long wave.

Periodic waves excited in the medium described by Eq. (3.27) are damped if their wavelength $\lambda < \lambda_0 = \sqrt{\gamma_2/\gamma_0}$, and grow amounting to finite amplitude for $\lambda > \lambda_0$. The amplitude of a stationary monochromatic wave can readily be estimated if its wavelength is only slightly greater than the critical value: $\lambda - \lambda_0 \ll \lambda_0$. In this case,

$$\eta = \sqrt{\frac{12}{\alpha\pi}\sqrt{\gamma_0^3\gamma_2(\lambda - \lambda_0)}},$$

i. e., the instability is saturated at the level $\eta \sim \sqrt{\lambda - \lambda_0}$.

Now consider the behaviour of long-wave perturbations with a spatial period $\lambda \gg \lambda_0$. The dissipative term $\gamma_2\partial^2\eta/\partial x^2$ in Eq. (3.27) may be neglected for these perturbations at the initial stage. The evolution of the long wave is then reduced to the growth of its amplitude under the action of long-wave instability (the term $\gamma_0\eta$ in (3.27)) and to simultaneous generation at higher harmonics, i. e., to energy transfer upwards the spectrum due to nonlinearity. As the wave grows, its profile is distorted, like the profile of a Riemann wave in an equilibrium medium, until steep fronts — "discontinuities" — are formed where short-wave dissipation is already significant. The energy absorption at the front, i. e., in the short-wave part of the spectrum, results in stabilization of instability and a stationary nonlinear wave is formed. Using a spectral language one can say that a constant energy flux is established over the spectrum. Assuming $\partial\eta/\partial t = 0$ yields the solution $\eta = \gamma_0 x$ that corresponds, like is the case in an equilibrium medium, to a saw-toothed wave (see Fig. 3.9a) with the amplitude

$$\eta \approx \frac{\gamma_0\lambda}{2\alpha}. \tag{3.28}$$

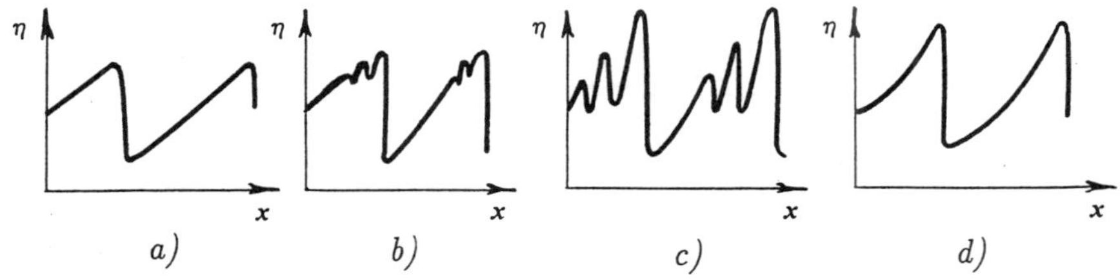

Figure 3.9: The shape of stationary waves: (a) in a medium with long-wave instability without dispersion; (b) at weak dispersion; (c) at strong dispersion; (d) in a medium with short-wave instability.

It should be emphasized that the amplitude of the established wave does not depend on the value of damping γ_2 which determines only the width of the wave front, that is, the thickness of the discontinuity. This thickness Δ may be estimated from the condition of balance between the nonlinear energy transport upwards the spectrum and the short-wave damping: $\alpha\eta\partial\eta/\partial x \sim \gamma_2\partial^2\eta/\partial x^2$. This, together with (3.28), gives

$$\Delta \sim \frac{2\gamma_2}{\lambda\gamma_0}. \tag{3.29}$$

Note that the saw-toothed wave with a period λ that appears in the evolution of initial sinusoidal perturbation is unstable relative to longer wave perturbations, which, in the final analysis, results in the onset of a wave with a maximum possible period. This period in the real ocean may be determined by both initial and boundary conditions.

The structure of the front can easily be found by considering the steady-state solutions of (3.27): $\eta = \eta(x - Vt)$. Bounded steady-state solutions exist only for $V = 0$. Integrating (3.27) for $\partial\eta/\partial t = 0$ yields

$$\alpha\frac{\eta^2 - \eta_0^2}{2} = \gamma_2\left(\frac{d\eta}{dx} - \gamma_0 \ln\frac{\gamma_0 - d\eta/dx}{\gamma_2}\right), \tag{3.30}$$

where η_0 is the parameter describing the amplitude. From (3.30) follows that the bounded steady-state solutions must be periodic, with their

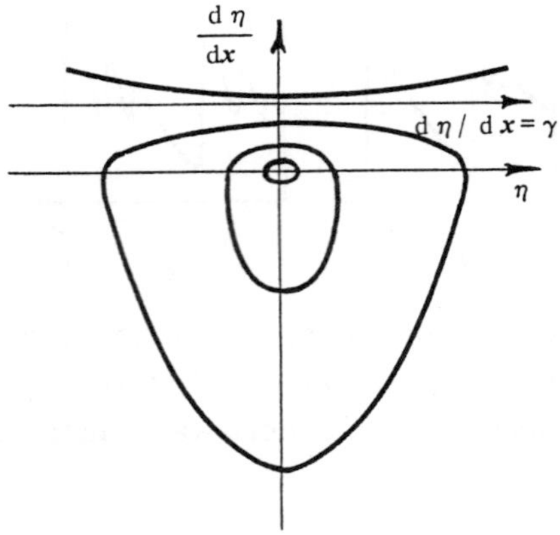

Figure 3.10: A phase plane for waves without dispersion in a medium with long-wave instability.

form approaching a saw-tooth, as η_0 grows. The phase portrait constructed from (3.30) is shown in Fig. 3.10. Large-amplitude waves have a slowly varying portion, to which the motion near the curve $d\eta/dx = \gamma_0$ corresponds, and a rapidly varying portion that corresponds to motion along the loop going far downwards.

So, in order to answer the question of how the dispersion affectsthe structure of discontinuity, one should modify Eq. (3.30) by introducing into it the term responsible for the high-frequency dispersion that enters Eq. (3.26) with the coefficient β:

$$\frac{\partial \eta}{\partial t} + \alpha\eta\frac{\partial \eta}{\partial x} + \beta\frac{\partial^3 \eta}{\partial x^3} = \gamma_0\eta + \gamma_2\frac{\partial^2 \eta}{\partial x^2}. \tag{3.31}$$

This equation is a generalization of the well-known Korteweg–de Vries–Burgers equation for nonequilibrium media (Ott *et al.*, 1973). The high-frequency dispersion, similarly to dissipation, does not affect the evolution of long waves until rather steep fronts are formed. At sufficiently strong dispersion ($\beta > \gamma_2^2/2\alpha\eta_0$), however, the front changes its shape on which damped oscillations appear (Fig. 3.9b).

If the dispersion is strong, so that the oscillations cut the entire wave profile (Fig. 3.9c), it is reasonable to speak not about the front and its structure but, instead, about the behaviour of the oscillations *per se* which transform into a train of solitary waves — solitons. This situation occurs at

$$\beta\frac{\partial^3\eta}{\partial x^3} \sim \alpha\eta\frac{\partial\eta}{\partial x} \gg \gamma_0\eta \sim \gamma_2\frac{\partial^2\eta}{\partial x^2}. \tag{3.32}$$

The soliton dynamics in a medium with weak nonlinearity and dissipation may be investigated in a more general case within the framework of Eq. (3.26). The shape of the soliton may be found from the Korteweg–de Vries equation that is derived from (3.31) for $\gamma_0 = \gamma_2 = 0$. Bearing in mind that the soliton parameters slowly vary at small $\gamma_{0,2}$, it is convenient to analyse nonstationary processes by the method of averaging over stationary waves (Pelinovsky, 1971; Ostrovsky, 1974; Whitham, 1974). The solution is represented in the form of a soliton with slowly varying parameters: amplitude A, width λ, and velocity V:

$$\eta = A(t)\,\mathrm{sech}^2\frac{(x - \int_0^t V(t)\,dt)}{\lambda(t)}, \tag{3.33}$$

where $V = \alpha A/3$ and $\lambda^2 = 12\beta/\alpha A$. From Eq. (3.26) follows the relation for the energy balance of waves that can readily be obtained by multiplying (3.26) by η and then integrating with respect to x:

$$\frac{d}{dt}\int_{-\infty}^{+\infty}\frac{\eta^2}{2}\,dx + \int_{-\infty}^{+\infty}\eta\hat{M}[\eta]\,dx = 0, \tag{3.34}$$

where $\hat{M}[\eta]$ is a dissipative operator corresponding to the right-hand side of (3.26). Substituting into (3.34) Eq. (3.33) we obtain an equation for the soliton amplitude

$$\frac{dA}{dt} = \frac{4}{3}\gamma_0 A + \frac{2.92}{\pi}\gamma_1\sqrt{\frac{\alpha}{12\beta}}A^{3/2} - \frac{4\alpha\gamma_3^2}{45\beta}A^2 - \frac{5.11\gamma_4}{\pi}\sqrt{\frac{\alpha}{12\beta}}A^{5/2}. \tag{3.35}$$

For Eq. (3.31), we have the coefficients $\gamma_1 = \gamma_3 = 0$. Then, the solution of (3.35) takes on a form

$$A(t) = \frac{A_0 A_\infty \exp(4\gamma_0 t/3)}{A_0[\exp(4\gamma_0 t/3) - 1] + A_\infty}. \tag{3.36}$$

The dynamics of solitons is quite simple: the solitons with the initial amplitude $A_0 < A_\infty$ are accelerated up to the velocity $V_\infty = \alpha A_\infty/3$ and are amplified up to the amplitude A_∞, while the solitons with $A_0 > A_\infty$ are slowed down and damped down to A_∞. The physics of this result is transparent: the small-amplitude solitons are broader and the long-wave instability leads to their amplification, while narrow (large-amplitude) solitons are damped due to short-wave dissipation. As a result, a regime in the form of solitons having equal amplitudes and velocities is established in the limit $t \to \infty$.

Numerical integration of Eq. (3.31) with periodic boundary conditions (in the "ring") was performed by Ostrovsky *et al.* (1985). Besides steady-state solutions — a periodic chain of solitons that is established in the evolution of initial conditions — periodic self-excited oscillations of the amplitudes of two solitons were also observed. The nonstationary bahaviour is explained by the fact that the soliton interaction in this case differs from that in a conservative system. The spectrum of the soliton pair narrows during collision and they acquire energy as a result of long-wave instability. Whereas, when the solitons exist separately and are spaced apart, they lose energy due to short-wave dissipation. The regime to be established depends on the length of the ring: if it is much longer than the characteristic size of the stationary soliton, then the "number of the degrees of freedom" is large, which allows for the existence of complex, probably, stochastic regimes of soliton interaction — soliton turbulence. Note that the stochastic solutions of an equation of the form (3.26) were actually found by Kuramoto and Tsuzuki (1976) and by Gasenko and Ivansky (1979).

3.4.2 Waves in media with short-wave instability

Wave propagation in an unstable medium occurs by a different scenario if the increment grows with the increase of the wave number (Fig. 3.8b). Such an instability, also referred to as "negative viscosity", is frequently realized in hydrodynamic flows. For example, for the waves on the surface of a viscous fluid flowing down a sloping plane, the instability increment is $\gamma(k) \sim k^2$ (Lin, 1974; Shen and Shin, 1974). In the absence of dispersion, the simplest equation for waves in a medium with short-wave instability has the following form:

$$\frac{\partial \eta}{\partial t} + \alpha \eta \frac{\partial \eta}{\partial x} + \gamma_2 \frac{\partial^2 \eta}{\partial x^2} = 0, \tag{3.37}$$

(where $\gamma_2 > 0$), i. e., it differs from the Burgers equation only by the sign of the coefficient γ_2 and transforms into it on the substitution

$$t \to -t, \quad x \to -x, \quad \eta \to -\eta.$$

Equation (3.37) allows for an exact solution: the substitution

$$\eta = \frac{2\gamma_2}{\alpha\theta} \frac{\partial \theta}{\partial x}$$

transforms it into a linear equation for heat conductivity:

$$\frac{\partial \theta}{\partial t} + \frac{\partial^2 \theta}{\partial x^2} = 0. \tag{3.38}$$

It can be shown (Pelinovsky and Friedman, 1974) that a smooth solution of (3.37) first evolves like a Riemann wave: its profile is getting steeper up to the formation of a discontinuity wave, with the discontinuity formed in some finite time — explosive instability of the progressive wave is realized. Using a spectral language this result may be explained as follows. As the short waves are amplified, their exponential growth is supplemented with the increasing energy flux from long-wave modes, which gives rise to the nonlinear increment of the short waves and provides conditions for explosion. It is interesting that the time of the "explosion" within (3.37) is equal to zero even for the sinusoid $\eta(0, x) \sim \sin(k_0 x)$.

Of course, no explosion occurs in a real situation. There occurs either nonlinear damping or a shorter-wave dissipation (the term proportional to γ_3 in a more complete equation (3.26)). Consequently, stationary waves having a characteristic pointed crest (Fig. 3.9d) are established. Long waves of this shape on a sheet were observed in the experiments described by Nakaryakov and Shreiber (1973). For shorter waves, the dissipation is significant ($\beta \neq 0$) in (3.26) and gives rise to oscillations in the wave profile which were observed on the surface of a sheet (P.Kapitsa and S.Kapitsa, 1949).

Let us now come back to the example that has already been considered above, i. e., to the evolution of long waves on a thermocline which is described by (3.26). In this case, the increment of radiative instability is $\gamma(k) \sim |k|$.[2] The nonlinear evolution of long waves gives rise, like for quadratic dependence of the increment $\gamma \sim k^2$, to long waves with a pointed crest; their shape was found numerically by Ott *et al.* (1973). The evolution of shorter waves for which dispersion is significant is completed by the formation of the solitons whose amplitudes meet Eq. (3.35) with the parameters $\gamma_0 < 0$, $\gamma_1 < 0$ and $\gamma_{2,3} > 0$.

The soliton dynamics is more diverse in this case. Equation (3.35) contains two equilibrium states that are determined by the roots in its right-hand side: A_1 and $A_2 > A_1$ (Fig. 3.11). If the initial soliton amplitude $A_0 < A_1$, the soliton damps. The solitons having a sufficiently large amplitude ($A_2 > A_0 > A_1$) are amplified up to $A = A_2$, while the solitons with the amplitude $A_0 > A_2$ are damped to the same steady-state value. Thus, hard excitation of solitons occurs: there exists the threshold value of the amplitude A_1 below which the solitons damp. The threshold of soliton generation is determined by turbulent dissipation (the coefficient γ_0 in (3.25)).

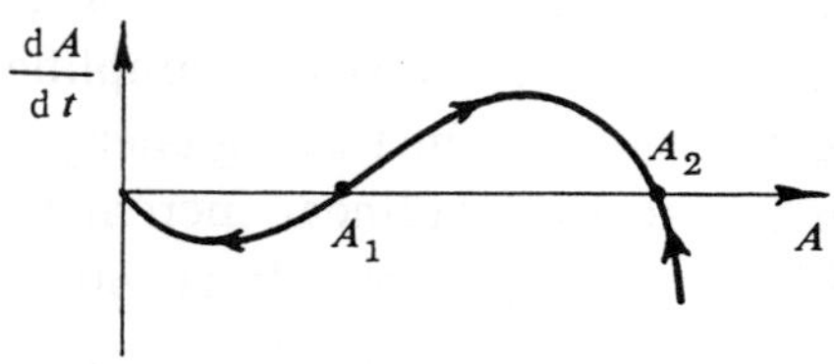

Figure 3.11: The curve on a phase plane corresponding to adiabatic evolution of solitons.

Note that for $A_1 \ll A_2$ the soliton with the amplitude $A_0 \approx A_1$

[2]A similar dependence on the wave number and the corresponding integral term in the equation in a single-wave approximation are known for the damping rate of Landau plasma waves (Pelinovsky, 1971; Karpman, 1975).

grows explosively at the initial stage:

$$A = \frac{A_0}{(1 - t/T_e)^2}, \quad \text{where} \quad T_e = \frac{2\pi}{1.46\gamma_1}\sqrt{\frac{3\beta}{\alpha A_0}}. \tag{3.39}$$

The explosive growth is restricted at the level $A = A_2$ thanks to short-wave dissipation.

3.5 Configuration instability of a vortex sheet in an incompressible fluid

Within the model of a vortex sheet in an homogeneous fluid, the Kelvin–Helmholtz instability increment grows without restriction, $\gamma(k) \sim K$, in the limit $k \to \infty$. Therefore, certain difficulties which need a more thorough analysis are often encountered in investigations of the evolution of initial perturbations localized in space. In particular, the shape of perturbations at large times strongly depends on the type of initial perturbation and on the rate of its spectrum decrease in the k-space. This phenomenon was called configuration instability (Nemtsov, 1986). Detailed investigation of the configuration instability of surface waves on a vortex sheet in an incompressible fluid, taking into account the surface tension at the interface between the moving layers, was performed by Nemtsov and Tsimting (1986). Results of their work will be presented below.

Suppose that a vortex sheet separates two layers of equal densities between which capillary forces act. Let the upper layer move to the right with the velocity $U/2$, and the lower layer move to the left with the velocity $-U/2$. Let us proceed from the linearized equations of motion for an incompressible fluid (3.1) with boundary conditions (3.2) and (3.3) (for $\rho_1 = \rho_2 = \rho$) supplementing them with initial conditions. For the formulation of the Cauchy–Poisson problem, one needs to know the initial displacement of the interface and its velocity. Assume for simplicity that displacement of the interface at the initial moment of time is known. Then, for $t = 0$ $(z = 0)$, we have

$$\begin{aligned} \xi_{1,2}(z=0) &= F(x), \\ \frac{\partial \xi_{1,2}}{\partial t}(z=0) &= 0. \end{aligned} \tag{3.40}$$

A dispersion equation for surface waves on a vortex sheet followsfrom Eq. (3.5) for $a = 1$:

$$\omega_{1,2} = \pm \frac{1}{2}\sqrt{2\sigma |k|^3/\rho - U^2k^2}. \tag{3.41}$$

The KH instability occupies the range of the wave numbers

$$0 < k < k_2 = \rho U^2/2\sigma,$$

with the maximal increment $\gamma_{\max} = \rho U^3/6\sqrt{3}\sigma$ attained at the point $k_{\max} = \rho U^2/3\sigma$ (Fig. 3.12). When $k > k_2$, there exist neutrally stable

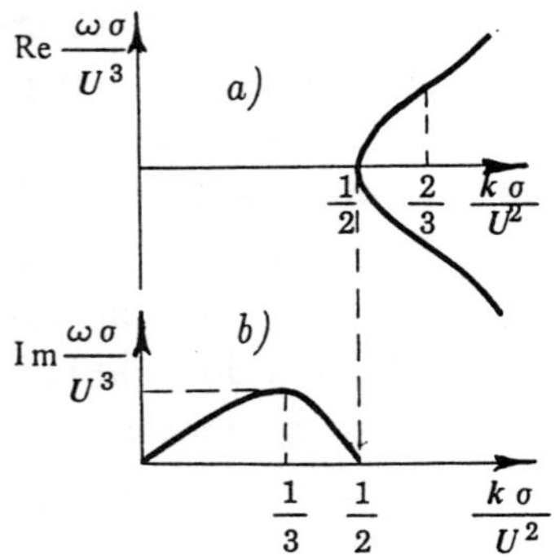

Figure 3.12: Real (a) and imaginary (b) branches of the dispersion curves for linear perturbations on a vortex sheet in a two-layer liquid of uniform density and with surface tension at the interface.

capillary waves. Let us represent the initial displacement of the interface in the form of the Fourier integral

$$F(x) = \frac{1}{2\pi}\int_{-\infty}^{+\infty} f(k)e^{ikx}\, dk. \tag{3.42}$$

Assuming the initial perturbation to be symmetric with respect to $x = 0$, one obtains a solution of the initial problem in the form

$$\xi_{1,2} = \frac{1}{\pi} \int_0^\infty f(k) \cos \omega_1 t \cos kx \, e^{ikz} dk. \tag{3.43}$$

This solution is symmetric relative to the $z = 0$ plane, therefore, below the integral (3.43) will be analysed only for $z > 0$.

Let us first ignore the surface tension ($\sigma = 0$), then formula (3.43) takes on the form

$$\xi_1 = \frac{1}{\pi} \int_0^\infty f(k) \cosh \frac{kUt}{2} \cos kx e^{kz} \, dk. \tag{3.44}$$

It is clear from (3.44) that for the solution to be bounded for any x and t, the spectrum of initial displacement at large k must decay faster than $\exp(-\alpha_0 k)$. Let us choose the spectrum of initial displacement in the form

$$f(k) = \exp(-\alpha_0 k - \alpha^n k^n), \tag{3.45}$$

where $n > 1$ is a natural number; in the calculations to follow it will be assumed that the inequality $\alpha \ll \alpha_0$ is met. The initial displacement will then have the form close to the Lorentz curve:

$$F(x) \approx \frac{1}{\pi} \frac{\alpha_0}{\alpha_0^2 + x^2}. \tag{3.46}$$

Analysis of (3.44) employing the method of steepest descents (Nemtsov and Tsimring, 1986) gives

$$\xi_1 = \frac{1}{8\pi\alpha} \begin{cases} e^q \sqrt{\dfrac{2\pi}{n(n-1)} \left(\dfrac{n}{\alpha}\right)^{\frac{n-2}{n-1}}}, & -\psi \le \arg \eta \le \psi, \\ -\dfrac{1}{\eta}, & \arg \eta \notin (-\psi, \psi). \end{cases} \tag{3.47}$$

Here, we designated

$$\eta = \frac{Ut/2 - z + ix - \alpha_0}{\alpha}, \quad q = (n-1)\left(\frac{\eta}{n}\right)^{\frac{n}{n-1}}, \quad \psi = \frac{\pi(n-1)}{2n}.$$

The region of space, where the relations (3.47) hold, is an isosceles ABC triangle with the apex angle B equal to 2ψ and the coordinates at the apex B being $(0, Ut/2)$ (see Fig. 3.13). As follows from (3.47), the field inside the ABC triangle grows exponentially with time:

$$\xi_1 \sim \exp\left[(n-1)\left(\frac{Ut - z + ix}{\alpha}\right)^{\frac{n}{n-1}}\right] + \text{c. c.}, \tag{3.48}$$

with the growth being unrestricted, as $\alpha \to 0$. This means that the KH instability incited by the initial displacement of the interface is developing inside a broadening ABC triangle.

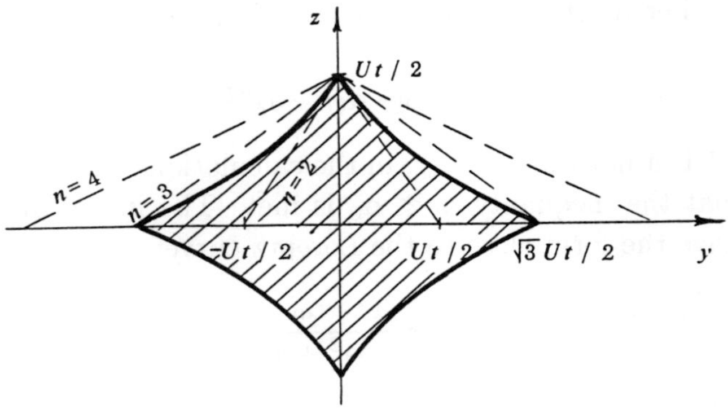

Figure 3.13: The region of space occupied by increasing perturbations on a vortex sheet in the absence (dashed curve) and in the presence (solid curve) of surface tension.

The relations obtained above show that the shape of the region of exponentially growing perturbations depends significantly on the high-frequency asymptotic behaviour of the spectrum of initial displacement (see (3.46) and (3.48)). Namely, as n grows, the apex angle B increases.

Although for $\alpha \ll \alpha_0$ the shape of the initial perturbation is close to the Lorentz curve and is weakly affected by the high-frequency part of the spectrum, the short-wave perturbations play a key role in the formation of the boundaries of instability region. Thus, we are concerned here with the configuration instability that is manifested in pronounced changes in the shape of the instability region with small variation of the initial conditions. It is also important that, besides exponential growth of the field in the instability region, the perturbations become small-scale (see (3.49)). In this connection, it is apparent that starting from some moment of time, the factors such as surface tension or viscosity restricting the KH instability at large k must be taken into account.

In order to take surface tension into account, let us again analyse the integral (3.43) but now for $\sigma \neq 0$. Capillarity will be insignificant at the initial stage of developing instability, if the source spectrum is localized in the long-wave range $k \ll \rho U^2/2\sigma$. For the spectra of the form (3.45), this corresponds to the condition $\alpha_0 \gg \alpha \gg \sigma/\rho U^2$. The growth of perturbations under such initial conditions at the first stage is accompanied by configuration instability: the value of the intergal (3.43) depends significantly on the particular short-wave asymptotic form of the spectrum $f(k)$. However, at sufficiently large $t \gg 2^{2-n}U^{2n-3}\alpha^n T^{1-n}$, the configuration instability vanishes. It should be noted that the absence of configuration instability does not indicate that the KH instability is stabilized. The perturbations inside the instability region continue to grow exponentially, but the shape of the instability region and the form of the solution no longer depend on the shape of initial perturbation.

The configuration instability is suppressed as follows. In passing over to the solutions whose configuration depends on the source spectrum to the stationary regime of KH instability development due to surface tension, the instability region broadens (in normalized variables $2z/Ut$, $2x/Ut$) for $n = 2$ and compresses for $n > 2$. The boundary of the region of exponentially large perturbations that is formed in the stationary regime is determined by the condition

$$\mathrm{Re}\,[\zeta^3 = (\zeta^2+3)^{3/2} - 9\zeta] = 0, \quad \text{where} \quad \zeta = 2(z + ix + \lambda_0)/Ut,$$

and lies between the triangle corresponding to $\sigma = 0$, $n = 2$ and the triangle corresponding to $\sigma = 0$, $n = 3$ (see Fig. 3.13). The boundary originates at the point $\zeta = 1$ ($z = Ut/2$) at an angle of 45° and is tangent to the x-axis at the point $\zeta = i\sqrt{3}$ ($x = Ut\sqrt{3}/2$). Figure 3.13 gives also the boundaries of the instability regions for $\sigma = 0$ when $n = 2, 3, 4$ which correspond to formula (3.48). In the center of the instability region ($|\zeta| \ll 1$), the perturbations grow by the law

$$\xi_{1,2} = \sqrt{\frac{\rho U}{\sigma t \pi \sqrt{27}}} \exp\left(\frac{\rho U^3 t}{6\sqrt{3}\sigma} \pm \frac{\rho U^2 z}{3\sigma}\right) \cos\left(\frac{\rho U^2 x}{3\sigma}\right). \tag{3.49}$$

This relation for $z = 0$ coincides to an accuracy of notation with the expression obtained by Shankar (1981).

Interesting phenomena occur also outside the instability region where there are no exponentially growing solutions. The shift of the boundary to the stability region is determined from the formula

$$\xi_1 \approx \frac{\rho U^2}{6\sigma\sqrt{\pi\delta}} \mathrm{Re}\, S, \tag{3.50}$$

$$S = \frac{\zeta + \sqrt{\zeta^2+3}}{\sqrt[4]{\zeta^2+3}} \exp\left\{\frac{2\delta}{27}\left[\zeta^3 + \sqrt{(\zeta^2+3)^3} - 9\zeta\right]\right\} + \frac{3}{\sqrt{\pi\delta}(\zeta+1)},$$

where $\delta = U^3 t/4\sigma$ is a large parameter and $*$ designates an operation of complex conjugation. For $\mathrm{Re}\,\zeta < 1/\delta \ll 1$, the first term in (3.51) has the order of magnitude $1/\sqrt{\delta}$, and the second one $1/\delta$. Therefore, the displacement near the interface depends only on the first term in (3.51) which corresponds to the excitation of capillary waves with large wave numbers (for $\mathrm{Im}\,\zeta = \sqrt{3}$ we have $k = 2\rho U^2/3\sigma$, and for $\mathrm{Im}\,\zeta \to \infty$ we get $k \to \infty$). Waves with small wave numbers lying in the interval $U^2/2T < k < 2U^2/3T$ are not excited in the evolution of initial localized perturbation in the region $x > Ut\sqrt{3}/2$.

This result looks quite unexpected because, according to the dispersion relation (3.41) (Fig. 3.12), in the region

$$U^2/2T < k < 2U^2/3T$$

there exist waves having group velocities equal to the group velocities in the short-wave part of the spectrum ($k > 2U^2/3T$). [3] At first sight, the wave field in the stability region must contain both short-wave and long-wave components (this usually occurs, for instance, when capillary-gravity waves are excited on a fluid surface). This is not the case here. According to the interpretation proposed by Nemtsov and Tsimring (1986), the contribution of the part of the spectrum belonging to the KH instability region ($0 < k < U^2/2T$) must be taken into account in the stability region too. The harmonics from this region form the perturbations which grow exponentially only in the instability region and do not grow outside it. The found solution demonstrates that the nongrowing "tails" of amplified waves interfere outside the instability region with the waves excited by the part of the spectrum that belongs to the stable region $k > U^2/2T$. As a result of this interference, the low-frequency part of the spectrum vanishes in the stability region.

Thus, although the dispersion curve (Fig. 3.13) indicates that at large x two classes of waves — in the range $U^2/2\sigma < k < 2U^2/3T$ and in the range $k > 2U^2/3T$ — must exist in different regions of the spatial spectrum, the localized source actually excites only waves in the second range mentioned. This result demonstrates that in analysing the evolution of initial perturbations in systems with instability, the dispersion curve should be employed with great caution. The example considered above illustrates that the intuition based on solutions of ordinary wave problems may give a wrong result.

The expression (3.51) shows also that the capillary waves are pressed against the interface. Their localization scale normalized to $Ut/2$:

$$\operatorname{Re}\zeta_0 = \frac{2z_0}{Ut} \approx \frac{9}{2\delta}\left(|\zeta|^2 + 3 + |\zeta|\sqrt{|\zeta|^2 - 3}\right)^{-1} \tag{3.51}$$

meets the inequality $\operatorname{Re}\zeta_0 \ll 1$. As the distance x increases at a given t, the wave field becomes more and more small-scale and is pressed against the interface $z = 0$. This is apparently caused by short capillary waves

[3] The group velocity of capillary waves on tangential velocity discontinuity is minimal for $k = 2U^2/3T$ and equals $c_g = U\sqrt{3}/2$.

which have larger group velocities and are the first to reach the point specified in space.

The configuration instability may be suppressed not only due to surface tension. Viscosity is another factor that restricts KH instability at large k and, as a consequence, destroys the configuration instability. If the viscosities ν are equal in both fluid layers, then a transition layer of thickness $d \sim \sqrt{\nu t}$ is formed at the interface and the perturbation having the scale $k^{-1} \sim \sqrt{\nu t}$ becomes stable (Betchov and Criminale, 1967). If the viscosity, ν_1, of one fluid is much greater than the viscosity, ν_2, of the other, then, as was mentioned above, the increment takes on a constant value $\gamma = U^2/2\nu_1$ for large $k \gg U/\nu_1$ in the limit $\nu_2 \to 0$ (Kikina, 1967). This may also lead to stabilization of configuration instability.

3.6 Nonstationary vortex sheet

We have so far considered wave processes in stationary shear flows. In real conditions, however, it is often necessary to take into account temporal variability of background hydrodynamic fields. The parametric effects occurring in these situations may be investigated on an example of simple shear flows in which temporal variability of the background is taken into consideration. Below, we will consider one of the simplest models – a vortex sheet with a nonstationary flow in an unbounded medium that was first analysed by Nesterov and Sekerge-Zen'kovich (1976).

Let us again take a model of a vortex sheet with density discontinuity and surface tension at the interface between layers (see Sect. 1.9 and Sect. 3.1), but now it will be assumed that the velocity of the upper layer is not constant; instead, it varies in time by the law $U(t)$, so that $U(0) = U_0 \neq 0$. While the small perturbations of the interface behave harmonically in time in the precritical regime in the absence of dissipative factors, now the function $\eta(x,t)$ may vary in a more complicated fashion. In particular, the amplitude of the perturbations may increase monotonically in time even in a precritical regime.

Let us find a particular form of the function $\eta(x,t)$ for different functions $U(t)$. The system (3.1), with the boundary conditions (3.2) and (3.3) taken into account, may give an equation for the amplitude $\tilde{\eta}(t)$ of the interface displacement: $\eta(x,t) = \tilde{\eta}(t)e^{ikx}$. This equation has the form:

$$(\rho_1+\rho_2)\frac{d^2\tilde{\eta}}{dt^2} + ik\rho_2 U(t)\frac{\partial\tilde{\eta}}{\partial t} + [gk(\rho_2-\rho_1) + \sigma k^3 \\ - \rho_1 k^2 U^2(t) + ik\rho_1 \dot{U}(t)]\tilde{\eta} = 0. \tag{3.52}$$

Substituting $\tilde{\eta} = Z(t)\exp\left[\frac{-ik\rho_1}{\rho_1+\rho_2}\int\limits_0^t U(t)\,dt\right]$ reduces the above equation to the form

$$\frac{d^2}{dt^2}Z(t) + \xi^2(t)Z(t) = 0, \tag{3.53}$$

where

$$\xi^2(t) = \frac{(\rho_2^2-\rho_1^2)gk + \sigma k^3(\rho_1+\rho_2) - \rho_1\rho_2 k^2 U^2(t)}{(\rho_1+\rho_2)^2}. \tag{3.54}$$

Let $U_0 < U_{KH}$, where

$$U_{KH} \equiv \sqrt{\left(\frac{1}{\rho_2}+\frac{1}{\rho_1}\right)\left[\frac{(\rho_2-\rho_1)g}{k} + \sigma k\right]}.$$

Designate $\omega_0 = \xi(0)$, where ω_0 is a real number. Let us seek solutions to Eq. (3.53) that would meet the initial conditions

$$Z(0) = A_0, \qquad \dot{Z}(0) = -i\omega_0 A_0. \tag{3.55}$$

These conditions mean that, if the flow velocity were constant and equal to U_0, then a harmonic wave having the amplitude A_0, frequency ω_0, and wave number k would propagate along the x-axes.

Assume first that $U(t)$ changes slower than the function $Z(t)$. Employing the method of averaging (Volosov, 1962), from (3.53) and (3.55) in a first approximation, one obtains

$$Z(t) = A_0\sqrt{\frac{\omega_0}{\xi(t)}}\exp\left[-i\int\limits_0^t \xi(t)\,dt\right]. \tag{3.56}$$

Then, for the interface η, we have

$$\eta(x,t) = A_0\sqrt{\frac{\omega_0}{\xi(t)}}\exp\left\{ikx - i\int\limits_0^t\left[\frac{akU(t)}{1+a} + \xi(t)\right]dt\right\}. \tag{3.57}$$

From (3.54) and (3.57) follows that, if the flow velocity grows in time, then the wave amplitude increases at the interface and the frequency diminishes even in the precritical regime.

Formulas (3.56) and (3.57) no longer hold in the neighbourhood of the critical point t_{cr} at which $\xi(t_{cr}) = 0$. It is possible, in principle, to find a solution by means of an asymptotic method in this case as well, but the solution will be too tedious.

Consider a flow having a variable velocity $U(t) = U_0\sqrt{1+\gamma t}$. Then $\xi(t) = \sqrt{\omega_0^2 - \mu t}$, where $\mu = \gamma a k^2 U_0^2/(1+a)^2$. The requirement of slow variation of $U(t)$ is given by the inequality $\kappa \equiv 2\omega_0/3\mu \gg 1$. In this case, for $Z(t)$, we have:

$$Z(t) = \sqrt{\frac{\pi\omega_0}{3\mu}}\exp\left[-i\left(\kappa - \frac{5\pi}{12}\right)\right]A_0 H_{1/3}^{(1)}\left(\frac{2\xi^3(t)}{3\mu}\right), \tag{3.58}$$

where $H_{1/3}^{(1)}$ is the Hankel function of order 1/3 of the first kind.

For the velocities far from the critical value, the argument of Hankel function $2\xi^3(t)/3\mu \gg 1$, therefore, instead of (3.58) we can write down its asymptotic form

$$Z(t) \approx \sqrt{\frac{\omega_0}{\xi(t)}}A_0\exp\left\{\frac{2i}{3\mu}[\xi^3(t) - \xi^3(0)]\right\}. \tag{3.59}$$

In the other limiting case, when the velocity $U(t)$ is close to the critical value, the inequality $2\xi^3(t)/3\mu \ll 1$ is fulfilled, therefore, for $Z(t)$ we have

$$Z(t) \approx \frac{i\Gamma(1/3)\sqrt{\omega_0}}{\sqrt{\pi}(3\mu)^{1/6}}A_0\exp\left[-i\left(\kappa - \frac{5\pi}{12}\right)\right]. \tag{3.60}$$

From this follows that the wave amplitude is amplified N times when the velocity tends to a critical value, with N being determined by the expression

$$N = \sqrt{\frac{\omega_0}{\pi}} \frac{\Gamma(1/3)}{(3\mu)^{1/6}} \approx 1.26 \left(\frac{\sqrt{1+a}}{\gamma a k^2 U_0^2} \right)^{1/6} \left[g(1-a) + \frac{\sigma}{\rho} k^2 - \frac{a U_0^2 k^{7/3}}{1+a} \right]. \tag{3.61}$$

The waves having the wave number

$$k_0 = \frac{\rho}{5\sigma(1+a)} \left[a U_0^2 + \sqrt{a^2 U_0^4 + 5 g \sigma (1-a^2)^2 \rho} \right] \tag{3.62}$$

have minimal amplification.

Let us now determine the time τ of the transition from the initial to the critical state, after which the KH instability is established for the given wave number k. From (3.54) one finds within the considered flow model that the transition time determined by the condition $\xi(t) = 0$:

$$\tau = \frac{\omega_0^2}{\mu} = \frac{1+a}{\gamma a U_0^2} \left[\frac{a U_0^2}{1+a} - \frac{g(1-a)}{k} - \frac{\sigma k}{\rho} \right] \tag{3.63}$$

is proportional to the square of the phase velocity of the wave. From this follows that the perturbations with $k_{KH} = \sqrt{\rho g(1-a)/\sigma}$ change to the critical state fastest of all (see Sect. 3.2). The minimal time needed to attain the critical state is equal to

$$\tau_{\min} = \frac{1}{\gamma} \left[1 - \frac{2(1+a)}{a U_0^2} \sqrt{\frac{g \sigma (1-a)}{\rho}} \right]. \tag{3.64}$$

There exist for the specified flow parameters U_0 and γ two waves having different k but equal phase velocities (in the gravity and capillary regions of the spectrum) which reach the critical state simultaneously. However, their amplification is different, from which follows that the accelerated flow selects modes by their wavelengths before transition toinstability.

To illustrate what was said above, consider the estimates correspon-ding to the vortex sheet at the water–air interface that were obtained by Nesterov and Sekerge-Zen'kovich (1976). Let us take $U_0 = 1$ m/s, $\gamma = 1\ \mathrm{s}^{-1}$, $\rho_2 = 1\ \mathrm{g/cm}^3$, $\rho_2/\rho_1 = 770$, and $\sigma/\rho = 74$ dyn cm^2/g. Then, for the wavelength $\lambda = 2$ cm ($k = 3.14\ \mathrm{cm}^{-1}$), from (3.61) one can find that the wave amplitude near the critical state is increased by 4.8 times during the time $\tau = 40$ s.

Let us consider now another model of a nonstationary flow – a vortex sheet with the velocity rapidly oscillating near the average nonzero value $U_0 \neq 0$, so that $U(t) = U_0(1 - \alpha \cos \Omega t)$. Supposing $\tau = \Omega t$, Eq. (3.53) may be transformed to the form

$$\frac{d^2 Z}{d\tau^2} + \left[\beta\varepsilon^2 - \varepsilon\left(\cos\tau + \frac{\alpha}{4}\cos 2\tau\right)\right] Z(\tau) = 0, \qquad (3.65)$$

where: $\varepsilon = 2\alpha\gamma_1 U_0^2/\Omega^2 \ll 1$,

$$\gamma_1 = \frac{ak^2}{(1+a)^2}, \qquad \omega_0^2 = \frac{(1-a)gk + \sigma k^3/\rho}{1+a} - \gamma_1 U_0^2\left(1 + \frac{\alpha^2}{2}\right).$$

Assume that the period of flow oscillations $2\pi/\Omega$ is small compared to the characteristic time of $Z(t)$ variation so that $\omega_0^2/\Omega^2 = \beta\varepsilon^2$, where β is a constant of order unity. Employing again the method of averaging over "fast variables" (Volosov, 1962; Bogolyubov and Mitropol'sky, 1963) one can easily find in a first approximation in ε the instability condition

$$\frac{\omega^2}{\gamma_1(1+\alpha^2/2)} < U_0^2 < \frac{\Omega^2(1+\alpha^2/2)}{2\gamma_1\alpha^2(1+\alpha^2/64)}, \qquad (3.66)$$

where $\omega^2 = [gk(1-a) + \sigma k^3/\rho]/(1+a)$ is an ordinary dispersion relation for capillary-gravity waves on water. Clearly, the instability occurs in a finite interval of U_0. Unlike the classical Kelvin–Helmholtz instability, stabilization in this case occurs at a rather large flow velocity U_0. The length of the interval of instability in U_0 depends on the amplitude of flow oscillations, α.

The minimal critical value $(U_0)_{\min}$ needed for the onset of instability in a rapidly oscillating flow is approximately $\sqrt{1+\alpha^2/2}$ times smaller than U_{KH} for a stationary flow. Thus, for example, while for the air flow above water surface $U_{KH} \approx 6.46$ m/s, $(U_0)_{\min} \approx 5.28$ m/s for $\alpha = 1$

and $(U_0)_{\min} \approx 3.73$ m/s for $\alpha = 2$. The estimates indicate that the pulsations of the wind may appreciably decrease the threshold of the emergence of KH instability and simplify the conditions of surface wave generation.

Let us now analyse the case when the flow oscillates in the absence of mean flow, i. e., suppose that

$$U(t) = U_1 \cos \Omega t. \tag{3.67}$$

Designating $\varepsilon = \gamma_1 U_1^2/8\Omega^2 \ll 1$ and $\beta\varepsilon = (2\omega^2 - \gamma_1 U_1^2)/8\Omega_1^2$, where $\beta = O(1)$, and employing the averaging method we find that primary motion is unstable if $\beta + 1/2 < 0$, which is equivalent to

$$\frac{4\omega^2}{\gamma_1}\left[1 + \sqrt{1 - \omega^2/2\Omega^2}\right]^{-1} < U_1^2 < \frac{4\omega^2}{\gamma_1}\left[1 - \sqrt{1 - \omega^2/2\Omega^2}\right]^{-1}. \tag{3.68}$$

One can see that here too the oscillation amplitude, U_1, has an upper and a lower limit. The inequalities (3.68) are simplified

$$\frac{2\omega^2}{\gamma_1} < U_1^2 < \frac{16\Omega^2}{\gamma_1} \tag{3.69}$$

under the condition $\omega^2 \ll \Omega^2$.

Now one can find the threshold value

$$(U_1)_{\min} = (1 + a)c_{\min}\sqrt{2/a},$$

where $c_{\min}$ is the minimal phase velocity of the waves at the interface of two layers at rest. For $a \ll 1$, $(U_1)_{\min}$ is much higher than $c_{\min}$, while for $a \approx 1$, $(U_1)_{\min} \approx 4c_{\min}$. For surface waves on water, we have $c_{\min} = 23$ cm/s and $(U_1)_{\min} \approx 10$ m/s.

Thus, in order to initiate the instability in a two-layer fluid without mean flows, quite substantial velocity oscillations of the upper layer are needed. Another conclusion that follows from (3.68) is that the oscillating flow may give rise to the instability of relatively low-frequency and, consequently, large-scale perturbations. There is no instability for $\omega > \Omega\sqrt{2}$. While for $\omega \approx \Omega\sqrt{2}$, the threshold value of the amplitude of flow pulsations needed for the excitation of perturbations with such a frequency is $\sqrt{2}$ times larger than $(U_1)_{\min}$.

In conclusion, we would like to note that no direct experiments were performed to verify the theoretical results presented above. However, the technique used by Thorpe (1984) for the investigation of KH instability in a tilted hydrochannel, where the tilting caused a relative motion of layers, was actually concerned with nonstationary (equally accelerated) motions of fluid.

3.7 Waves on vorticity discontinuity

Besides the model of a vortex sheet, a model with a piecewise-linear velocity profile is frequently employed in hydrodynamic theory of stability as well as in other problems of hydrodynamics. Within this model the velocity profile is continuous, which makes the model more realistic for practical use and the vorticity, being constant in all layers with a linear velocity profile, makes a jump at the layer interface. Using piecewise-linear models one can approximate an arbitrary continuous velocity profile. This approach, which goes back to Rayleigh (1945), is known as the algebraic Rayleigh's method. In this case, however, there often arise the questions associated with the convergence of the method and with the relation of dispersion characteristics for continuous and piecewise-linear velocity profiles. The first question was answered by Gertsenshtein (1969) where the convergence of Rayleigh's method for unrestricted increase of the number of bends, n, in the velocity profile was justified rigorously. As n grows, the dispersion curves of the piecewise-linear model transform to the dispersion curves found from the boundary-value problem for the Rayleigh equation (see Chap. 2). For a small amount of bends, the dispersion curves coincide qualitatively and quantitatively only for the first modes in the region of rather long waves. This is, in part, an answer to the second question.

Another peculiarity of piecewise-linear velocity profiles is that they can give a simple physical interpretation of some results available in the theory of hydrodynamic stability which were earlier obtained formally

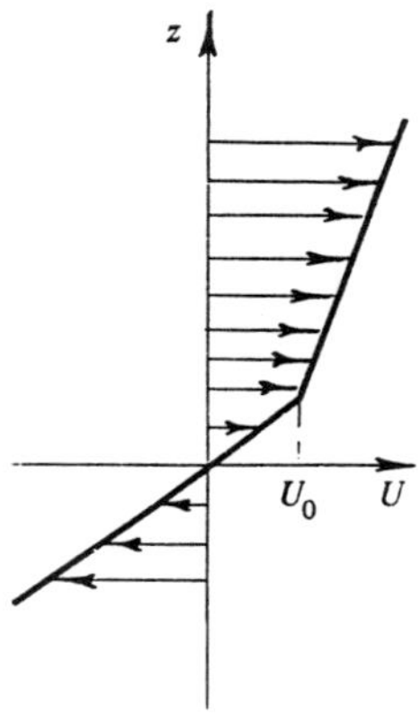

Figure 3.14: A model of a shear flow with a single bend in the velocity profile.

mathematically.

Consider first a flow with a piecewise-linear velocity profile that has only one bend (see Fig. 3.14). It should be emphasized that the flow having such a profile is of separate interest because it gives a good approximation of some flows in real conditions and, besides, it can be readily reproduced in laboratory conditions. Homogeneous flows having such a velocity profile may contain waves at vorticity discontinuity, the so-called "flow waves" with a simple dispersion relation: $\omega = kU_0$, where U_0 is the velocity of the flow at the inflection point of the profile (Fig. 3.14). Direct calculations readily verify (Ezersky *et al.*, 1981) that the energy of the flow wave in a linear approximation is determined by the expression

$$E = (U_2' - U_1')U_0|\eta|^2, \tag{3.70}$$

where U_1' and U_2' are the derivatives of the velocity profile under and above the bend, respectively (i. e., the vorticities in each layer); and η is the displacement of the layer interface. One can see from (3.70) that the wave perturbations may have both positive and negative energy. Negative energy waves (cf. also Sect.1.8) may give rise to instability in the presence of dissipative mechanisms.

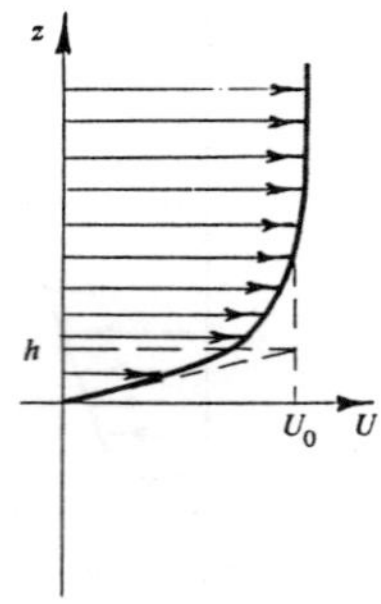

Figure 3.15: A model of a wind profile above water surface (solid curve) and its piecewise-linear approximation (dashed curve).

A pictorial example of using a piecewise-linear model of a flow is, again, the problem of wind generation of surface waves on water (Ezersky *et al.*, 1981). The profile of the wind and its piecewise-linear approximation are depicted in Fig. 3.15.

This model contains two modes: a surface gravity wave pressed against density discontinuity and a flow wave pressed against vorticity discontinuity at height h above water. A complete dispersion equation describing the two modes has the following form

$$\left[\omega - kU_0 + \frac{U_0}{2h}\left(1 - e^{-2kh}\right)\right]\left[\omega^2 - \frac{(\rho_2 - \rho_1)gk + \sigma k^3 + U_0\omega\rho_1/h}{\rho_1 + \rho_2}\right]$$

$$= -\frac{\rho_1\omega^2 U_0 e^{-2kh}}{h(\rho_1 + \rho_2)}. \qquad (3.71)$$

When the regions of wave localization are spaced apart ($h \to \infty$), this dispersion equation is divided into two independent equations: for the surface capillary-gravity wave

$$(\rho_1 + \rho_2)\omega^2 - (\rho_2 - \rho_1)gk - \sigma k^3 = 0$$

and for the flow wave

$$\omega - kU_0 = 0.$$

In the case of interest ($U_2' < U_1'$), the flow wave has negative energy. While the surface gravity wave remains positive because the velocity shift actually does not affect this wave due to small air density.

Until the flow velocity U_0 is sufficiently small, the dispersion branches of the surface and the flow waves do not intersect. However, as U_0 increases, starting from some value U_c, these branches begin to reconnect. With the wave coupling (the right-hand side in Eq. (3.71)) taken into account, the dispersion curves reconnect. The rule of reconnection is determined, as was mentioned in Sect. 1.8, by the signs of the energies of the corresponding waves. In our case, these signs are equal, which gives rise to the formation of instability regions in the neighbourhood of intersection points (see Fig. 3.16).

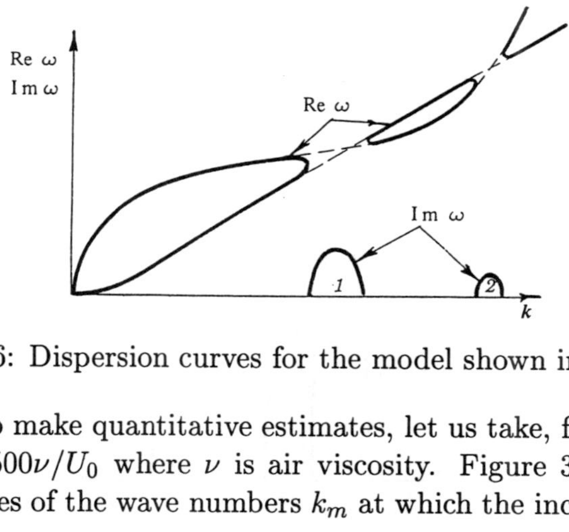

Figure 3.16: Dispersion curves for the model shown in Fig. 3.15.

In order to make quantitative estimates, let us take, following Chan (1977), $h = 500\nu/U_0$ where ν is air viscosity. Figure 3.17 shows the U_0-dependences of the wave numbers k_m at which the increment has an extremum.

With capillarity taken into account, there are two instability regions. Accordingly, the curve $k_m(U_0)$ consists of two branches which merge when $U_0 = U_c$, where

$$U_c \equiv \sqrt[4]{\frac{4(\rho_2 - \rho_1)g\sigma}{(\rho_2 + \rho_1)^2}}.$$

The critical value of the air flow velocity $U_c = 23.3$ cm/s corresponds

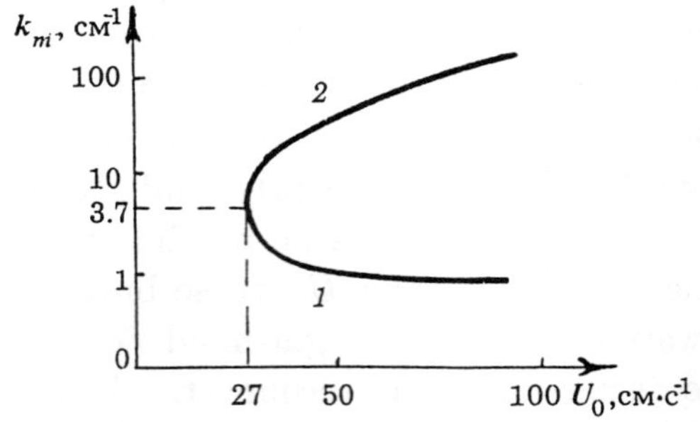

Figure 3.17: The dependences of wave numbers on flow velocity for which the growth rate is maximal in regions 1 and 2 marked in Fig. 3.16.

to the threshold of generation of the surface waves having an identical phase velocity and the wavelength $\lambda = 1.7$ cm.

The increments are plotted as a function of the wave number k in Fig. 3.18 for several values of U_0. One can see from the plots, that

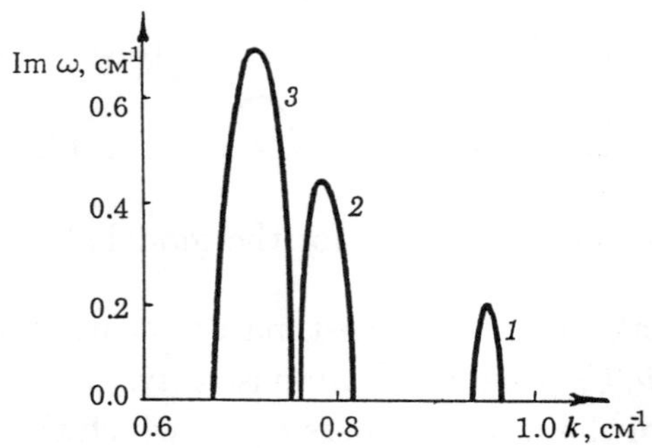

Figure 3.18: Growth rate versus wave number for the gravity branch (region 1 in Fig. 3.16) at different values of U_0: (1) – $U_0 = 0.5$ m/s; (2) – $U_0 = 0.7$ m/s; (3) – $U_0 = 0.9$ m/s.

as U_0 increases, the instability region is expanded and shifted to the region of small k and the maximal increment grows. In a weak coupling approximation, it is easy to obtain an expression for the growth rate of waves and the width of the interval Δk corresponding to the growing perturbations. The increment is maximal at the points

$$(k_m)_{1,2} = \frac{U_0^2(\rho_2+\rho_1)}{2\sigma}\left[1 \pm \sqrt{1 - \frac{4\sigma g(\rho_2-\rho_1)}{(\rho_2+\rho_1)^2 U_0^4}}\right]. \tag{3.72}$$

The maximal increment $\gamma_{\max}$ and Δk are then expressed by

$$\gamma_{\max} = e^{-k_m h}\sqrt{\frac{\rho_1}{\rho_1+\rho_2}\frac{U_0^2 k_m}{2h}}; \qquad \Delta k = \frac{2\gamma_{\max}}{U_0 - c_g}, \tag{3.73}$$

where c_g is the group velocity of the surface wave at the point of intersection with the flow wave (for $k = k_m$). The results obtained from these formulas agree well with data of direct numerical computation using the expression (3.71).

Note that the considered instability, that arises in flows with piecewise-linear velocity profiles, is related to Miles' instability for flows with smooth velocity profiles (Miles, 1957b) similarly to the hydrodynamic and kinetic instabilities in a plasma. The instability considered here is a more rough model that is based on the approximation of a real smooth vorticity profile by a stepwise (piecewise-constant) function.

3.7.1 Algebraic Rayleigh's method. Interpretation of hydrodynamic instability

A piecewise-linear model may also be a useful tool for the interpretation of the criterion of wind instability of surface waves obtained by Miles (1957b). This criterion (see Sect. 2.2) consists in the following. The surface waves are unstable, if the wind velocity profile $U(y)$ has a positive curvature $d^2U/dy^2 < 0$ in the critical layer where the wind velocity is equal to the phase velocity of the wave. It is clear from (3.71) that the sign of the energy for the flow wave in a model with a bended velocity profile depends on the sign of the difference $U_2' - U_1'$, that is, the analog of the second derivative d^2U/dy^2. For $d^2U/dy^2 < 0$, the waves possess negative energy and there arises instability; while for $d^2U/dy^2 > 0$ (for example, for the wind profile depicted in Fig. 3.19) no instability occurs because the flow waves and the gravity wave have energies of equal signs.

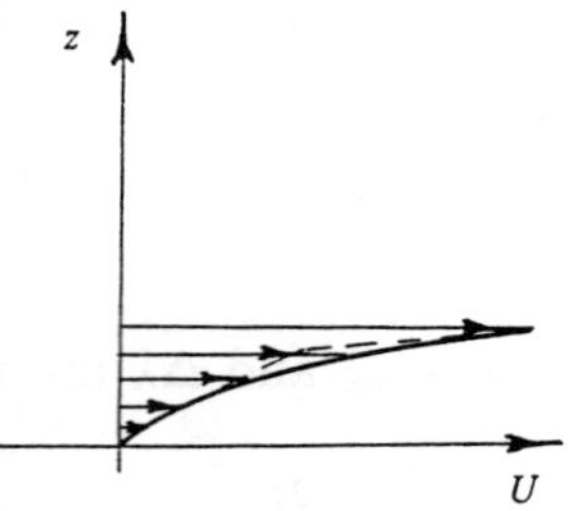

Figure 3.19: The model of wind profile above water surface (solid curve) within which surface waves are not generated; its piecewise-linear approximation (dashed curve).

The corresponding reconnection of dispersion curves is demonstrated in Fig. 3.20.

Now consider shear flows with smooth profiles (Fig. 3.21) and approximate the profiles by piecewise-linear curves in accordance with the algebraic method. The amount of flow waves in this model equals the number of bends. All the flow waves here have energies of equal signs: either positive (Fig. 3.21a) or negative ones (Fig. 3.21b). Therefore, even when the dispersion curves corresponding to these modes approach one another, they reconnect as they would in a stable case (see Fig. 3.20).

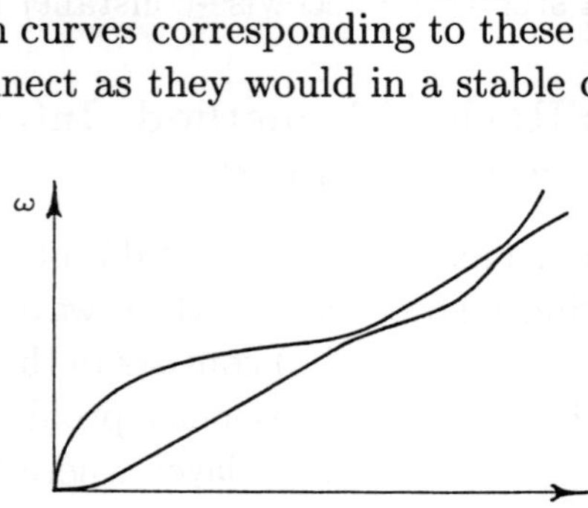

Figure 3.20: Reconnection of the dispersion curves corresponding to the model given in Fig. 3.19.

This fact is in conformity with Rayleigh's theorem according to which flows without inflection points in the profile $U(z)$ are stable in a homogeneous ideal fluid. Thus, the onset of instability in the presence

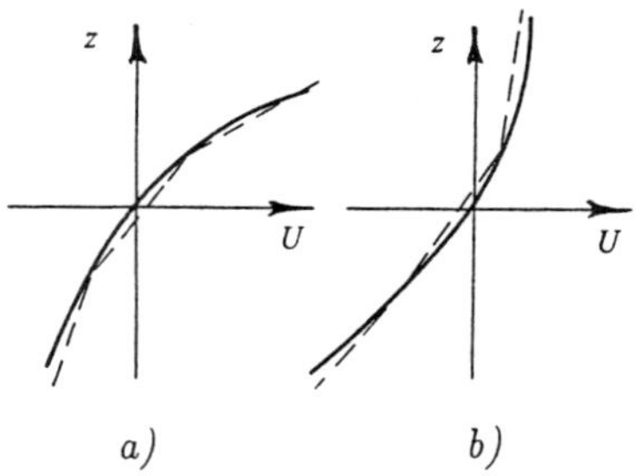

Figure 3.21: Two types of flow profiles without inflection points (solid curves) and their piecewise-linear approximations (dashed curves).

of inflection points may be explained by the interaction of flow waves having energies of different signs.

Another example allows one to specify Rayleigh's theorem and interprete its generalization (the Fjørtøft–Høiland theorem (Fjørtøft, 1950; Høiland, 1953)). According to the results obtained by Fjørtøft and Høiland, the instability is implemented in a homogeneous ideal fluid, if the vorticity profile $U'(z)$ has not merely an extremum but necessarily a maximum at the inflection point. Flows with a vorticity minimum at the inflection point are stable. These conclusions, following from the integral relations that are derived from Rayleigh's equation (2.1), are illustrated by a simple example. Let us take shear flows whose profiles are depicted in Fig. 3.22.

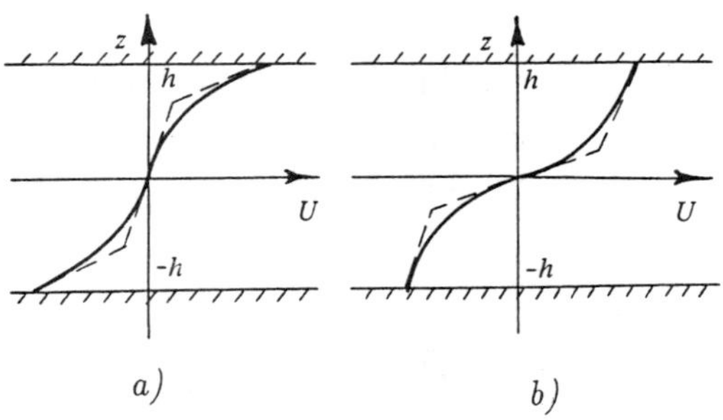

Figure 3.22: Two types of flow profiles with inflection points (solid curves) and their piecewise-linear approximations (dashed curves).

Approximating them by piecewise-linear broken functions with two bends, one can write them in the form

$$U(z) = \begin{cases} \Omega_+ z + (\Omega_0 - \Omega_+)h, & z \geq h; \\ \Omega_0 z, & |z| \leq h; \\ \Omega_- z + (\Omega_- - \Omega_0)h, & z \leq -h. \end{cases} \tag{3.74}$$

Figure 3.22 and the formula (3.70) indicate that the wave modes pressed against the layers of vorticity discontinuity have different signs of energy: the perturbations corresponding to the upper bend in Fig. 3.22a have positive energy and the perturbations corresponding to the lower bend have negative energy. In Fig. 3.22b, on the contrary, the perturbations at the lower bend possess positive energy and those at the upper bend have negative energy. When the dispersion curves corresponding to these perturbations intersect, instability may be expected. It is a rather simple task to calculate dispersion curves in these models. From Rayleigh's equation (2.1) (which is trivial: $\psi'' = k^2\psi$ for piecewise-linear flows) follows that the solutions in each layer are expressed through the exponents:

$$\psi = \begin{cases} Ae^{kz}, & z \leq -h; \\ Be^{kz} + Ce^{-kz}, & |z| \leq h; \\ De^{-kz}, & z \geq h. \end{cases} \tag{3.75}$$

In order to match the solutions at the layer interface, ordinary boundary conditions — discontinuity of pressure and of vertical velocity component — will be employed, which is equivalent to the following equalities:

$$\psi_+ = \psi_- \quad \text{and} \quad \left(\psi'_+ + \frac{U'_+}{c - U_+}\psi\right)_+ = \left(\psi'_- + \frac{U'_-}{c - U_-}\psi\right)_- . \tag{3.76}$$

The subscripts "+" and "−" refer to the correspondingquantities above and below the velocity bend.

The conditions (3.76) together with (3.75) yield a dispersion equation

$$\omega_{1,2} = \frac{\Omega_+ - \Omega_-}{4} \pm \frac{1}{4}\left[(\Omega_+ - \Omega_-)^2 - 8U_0k(2\Omega_0 - \Omega_+ - \Omega_-)\right.$$

$$\left. + 16U_0^2k^2 + 8(\Omega_0 - \Omega_-)(\Omega_0 - \Omega_+)\frac{\tanh 2kh}{1+\tanh 2kh}\right]^{1/2}, \qquad (3.77)$$

where $U_0 \equiv \Omega_0 h$ is the flow velocity at the upper inflection point.

It is assumed in the analysis of Eq. (3.77) that the width of the mean layer, h, is very large, so that for perturbations of arbitrary scale, except a very narrow region near $k = 0$, it can be supposed to a high accuracy that $\tanh 2kh \approx 1$. Then, the expression (3.77) takes on a simpler form:

$$\omega_1 = U_0k - \frac{\Omega_0 - \Omega_+}{2}; \qquad \omega_2 = -U_0k + \frac{\Omega_0 - \Omega_-}{2}. \qquad (3.78)$$

One can easily see that the dispersion curves (3.78) do not intersect for the flows depicted in Fig. 3.22a. Allowance made for the coupling between two wave modes (i. e., the finiteness of the mean layer thickness) only slightly modifies these dispersion curves. Therefore, the flows having such profiles are stable in a homogeneous ideal fluid in accordance with the Fjørtøft–Høiland theorem (the profile of a flow withminimal vorticity).

In the presence of stratification, the flows with minimal vorticity may also become unstable because negative energy waves may now interact with internal positive energy waves. The flow waves at the vorticity discontinuity give up their energy to internal waves and increase in conformity with what was said in Sect. 1.8. An example of such an instability for flows with a smooth profile was considered analytically by Thorpe (1969). In this context, it is easy to understand the loss of stability in a shear flow with introduction of statically stable stratification which might appear to be a paradox at first sight (see, for example, the book of Turner (1973)). The flow instability in a boundary layer in the presence of small viscosity that was revealed by Heisenberg back in the 1920's (Lin, 1955) may be interpreted analogously.

The situation is quite different for the flow profiles presented in Fig. 3.22b. The dispersion straight lines (3.78) intersect, which actually indicates the reconnection of true dispersion curves in the given system, with the coupling between wave modes taken into account. Since the intersecting straight lines correspond to the waves having energies of different signs, the reconnection occurs as it would in unstable systems (see Sect. 1.8), i. e., imaginary ω correspond to real k in a definite range of the wave numbers. This explains the presence of instability in the flows possessing a vorticity maximum at the inflection point of the profile.

Thus, the concept of negative energy waves is a useful tool for interpretation and prediction of instabilities in shear flows. Some results that were earlier obtained formally become physically meaningful. Besides, this concept is heuristic to a certain extent. It can be used to predict the appearance of new wave effects in hydrodynamic flows.

Chapter 4

Over-reflection

In this chapter we consider different aspects of one of the most effective mechanisms of wave amplification and absorption that is associated with the presence of negative energy waves in a hydrodynamic flow. It is convenient to analyse the characteristic features of this mechanism on the examples of simple hydrodynamic flows: a vortex sheet (or a tangential discontinuity of flow velocity) and other flows having piecewise constant vorticity in which the effect of amplification is manifested in a pure form and allows for analytical investigation. Such flows can be considered as models for some realistic flows in hydrodynamics, in physics of atmosphere, in oceanic physics, plasma physics, magnetohydrodynamics, etc. The effect of wave amplification at the reflection from shear flow (over-reflection) was first revealed for the acoustic waves incident on a vortex sheet (Miles, 1957a, 1958) and was later considered for other types of waves in moving media, such as internal gravity waves (McKenzie, 1972), magnetohydrodynamic waves (Feijer, 1963), electromagnetic waves (Lupanov, 1975) and others.

4.1 Miles–Ribner problem

Consider first in detail a simple problem of monochromatic, acoustic wave reflection at a vortex sheet (Fig. 4.1). The field of the harmonic

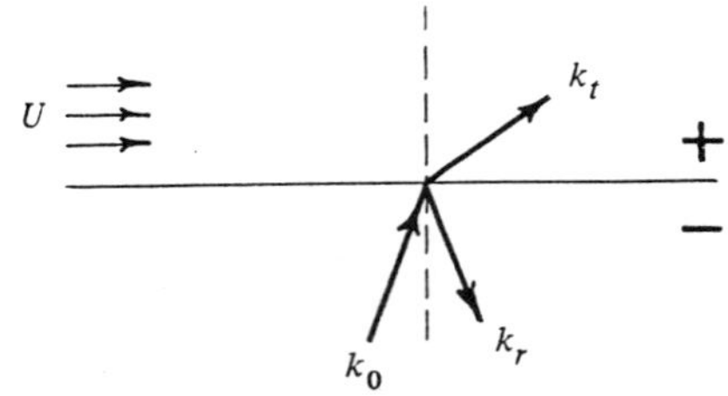

Figure 4.1: The reflection and refraction of a sound wave at a vortex sheet.

disturbances proportional to $\exp(-i\omega t + ikx)$ for potential φ, pressure p and the y-component of displacement ζ of particles in moving (the subscript "+") media and in media at rest (the subscript "−") may be represented in the form:

$$\begin{aligned} \varphi_- &= e^{iq_-y} + \mathrm{R}e^{-iq_-y}, & \varphi_+ &= Te^{iq_+y}, \\ p_- &= i\omega\rho\varphi_-, & p_+ &= i(\omega - kU)\rho\varphi_+, \\ \zeta_- &= -\frac{q_-}{\omega}(e^{iq_-y} - \mathrm{R}e^{-iq_-y}), & \zeta_+ &= -\frac{q_+}{\omega - kU}Te^{iq_+y}, \end{aligned} \tag{4.1}$$

where R and T are complex reflection and transmission coefficients. They can readily be found by matching solutions in both media on the surface $y = 0$ using boundary conditions $p_- = p_+$ and $\zeta_- = \zeta_+$:

$$R = \frac{q_-/\omega^2 - q_+/(\omega - kU)^2}{q_-/\omega^2 + q_+/(\omega - kU)^2}, \quad T = \frac{2q_-/[\omega(\omega - kU)]}{q_-/\omega^2 + q_+/(\omega - kU)^2}, \tag{4.2}$$

where $q_- = \sqrt{\omega^2/c_s^2 - k^2}$, $q_+ = \sqrt{(\omega - kU)^2/c_s^2 - k^2}$, and c_s stands for sound velocity. The sign of the y-component, q_+, of the wave vector in a moving medium depends on the radiation condition $(c_g)_y > 0$ that may be found from the solution of a boundary value problem (Miles, 1958). From the dispersion equation

$$(\omega - kU)^2 = c_s^2\,(k^2 + q_+^2)$$

follows

$$(c_g)_y \equiv \frac{\partial \omega}{\partial q_+} = \frac{c_s^2\, q_+}{\omega - kU}. \tag{4.3}$$

The radiation condition demands that the branch $q_+ < 0$ should be chosen for $\omega - kU < 0$. The reflected wave is amplified in this case: $|R| > 1$.

Interpretation of the effect of over-reflection is associated with finding the sign of the refracted wave energy. The energy and momentum of a sound wave in a medium at rest are expressed through the formulas (1.31). Taking into consideration that the pressure amplitude in a moving medium is independent of the reference system, (1.31), with (1.4) taken into account, yields average energy density:

$$E = E_0 \frac{\omega}{\omega - kU} = \frac{\omega |p|^2}{2\rho c_s^2(\omega - kU)}. \tag{4.4}$$

The energy density is negative for $\omega - kU < 0$. The vertical component of energy density flux is then

$$S_y = (c_g)_y E = \frac{q_+ \omega |p|^2}{2\rho(\omega - kU)^2} = \frac{1}{2}\rho\omega q_+ |\varphi|^2. \tag{4.5}$$

When $q_+ < 0$, the energy flux of the refracted wave is directed towards the vortex sheet: $S_y < 0$. Thus, amplification occurs due to the energy contribution from the moving medium, with the NEW escaping to the moving medium. The energy conservation law:

$$q_-(1 - |R|^2) = q_+ |T|^2$$

may be verified directly from the expressions (4.2).

Consider different regimes of reflection depending on the angle of incidence θ and Mach number $\mathrm{M} \equiv U/c_s$. Representing the wave vector of the incident wave in the form $\mathbf{k}_0 = (\omega/c_s)(\sin\theta, \cos\theta)$ the reflection coefficient (4.2) may be written as

$$R = \frac{\cos\theta - \sqrt{1 - \sin^2\theta/(1 - \mathrm{M}\sin\theta)^2}}{\cos\theta + \sqrt{1 - \sin^2\theta/(1 - \mathrm{M}\sin\theta)^2}}. \tag{4.6}$$

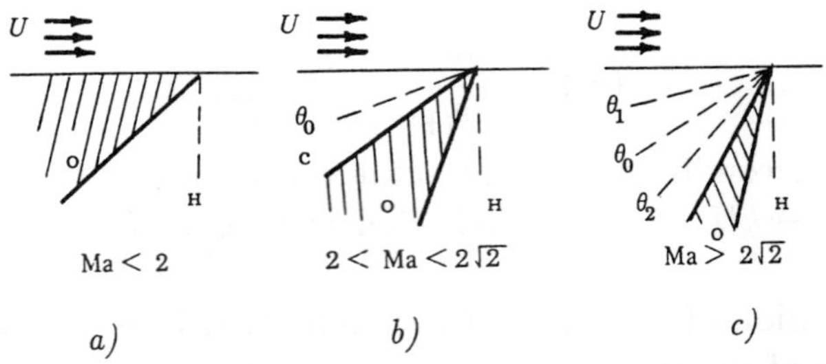

Figure 4.2: Different regimes of sound reflection at a vortex sheet. The ranges of the angles of incidence are designated through H (normal incidence), *o* (total or full reflection, hatched), and *c* (over-reflection).

Three different regimes of reflection are possible (Fig. 4.2):
(i) Normal reflection ($q_+ > 0, \ |R| < 1$) for $\sin\theta < 1/(\mathrm{M}+1)$.
(ii) Full reflection ($\mathrm{Re}\, q_+ = 0, \ |R| = 1$) for $1/(\mathrm{M}+1) \leq \sin\theta \leq 1$, if $\mathrm{M} \leq 2$ and for $1/(\mathrm{M}+1) \leq \sin\theta \leq 1/(\mathrm{M}-1)$, if $\mathrm{M} > 2$.
(iii) Over-reflection ($q_+ < 0, \ |R| > 1$) for $\mathrm{M} > 2$ and $\sin\theta > 1/(\mathrm{M}-1)$.

In the latter case there exists a resonance angle of incidence

$$\theta_0 = \arcsin(2/\mathrm{M}) \tag{4.7}$$

for which $|R| = \infty$. Spontaneous "Cherenkov" radiation of a vortex sheet, i. e., an infinitely narrow liquid layer moving with a velocity $U/2$ where vorticity $\Omega \equiv \mathrm{rot}\, v$ is nonzero (in other words velocity discontinuity on a vortex sheet) occurs under this angle. A negative energy wave is radiated into a moving medium and a positive energy wave into a medium at rest in the course of spontaneous radiation. Oscillations of the vortex sheet are neither damped nor amplified and the energy of the sound radiated into the medium at rest is drawn from the entire moving medium.

When $\mathrm{M} \geq 2\sqrt{2}$, the vortex sheet becomes stable (Landau and Lifshits, 1986) while unstable surface Kelvin–Helmholtz modes are transformed into waves travelling at the vortex sheet with a phase velocity

$$c = \pm c_s \left[1 + \mathrm{M}^2/4 - \sqrt{1+\mathrm{M}}\right]. \tag{4.8}$$

By virtue of $c + U/2 > c_s$, "Cherenkov" radiation must arise. The perturbations travelling over the vortex sheet radiate a sound at the angles $\theta_{1,2}$ for which

$$\sin\theta_{1,2} = \left[\mathrm{M}/2 \pm \left(1 + \mathrm{M}^2/4 - \sqrt{1+\mathrm{M}}\right)\right]^{-1}. \tag{4.9}$$

A vortex sheet in a compressible gas is stabilized, for $\mathrm{M} \geq 2\sqrt{2}$, only within a two-dimensional model. The projection of the flow velocity onto the wave vector direction, that determines the action of the medium motion on wave propagation (which is, actually, the Dopler shift), decreases for the perturbations travelling at an angle $\tilde{\theta}$ to the plane depicted in Fig. 4.1. Consequently, the condition of stability of the perturbations takes on the form

$$\mathrm{M}\,\cos\tilde{\theta} \geq 2\sqrt{2}. \tag{4.10}$$

Thus, the perturbations travelling at a sufficiently large angle $\tilde{\theta}$ are always unstable, i. e., total stabilization of a supersonic vortex sheet is impossible in gas (Syrovatsky, 1954). Taking into consideration that the wave is no longer "sensitive" to the motion of the medium and the instability vanishes for $\tilde{\theta} = \pi/2$, one arrives at a conclusion that there exist maximally stable perturbations for which the value $\mathrm{Im}\,(\omega/k)$ and the corresponding angle θ_{max} depend on Mach number M. Those dependences were calculated by Kikina and Sannikov (1969).

Note, however, that the solutions obtained within a model of two-dimensional gas dynamics may also be realized for surface waves in a shallow water approximation (Landau and Lifshits, 1986) where, in particular, a vortex sheet may be stabilized (Nezlin and Snezhkin, 1990).

NEWs in a supersonic vortex sheet may give rise to various dissipative instabilities. For instance, a sound reflecting boundary placed in a moving medium supplements the vortex sheet which amplifies the reflected sound with acoustic feedback (Kolykhalov, 1984; Acheson, 1986). The arising instability can then be interpreted as dissipative instability of modes of a waveguide formed by the vortex sheet and the reflecting boundary. It can be shown that these modes possess negative energy and dissipation occurs due to acoustic radiation into medium at rest. Thus, an acoustic variant of the radiative instability considered in Sect. 3.3 emerges in this case. Such an instability was observed in supersonic boundary layers (Betchov and Criminale, 1967).

Instability also arises when a vortex sheet is replaced by a transition layer of finite thickness (Blumen *et al.*, 1975). The poles of the reflection coefficient $R(\omega, k)$ leave the real axis ω for the complex plane (Graham, B. and Graham, E., 1969). The neutrally stable normal modes of the vortex sheet that are accompanied by spontaneous sound radiation are then transformed to unstable modes.

The problem of reflection of a plane monochromatic wave may be incorrect for an unstable vortex sheet. Legitimacy of solutions must be verified in the framework of more complex models that would take into account variability and inhomogeneity of the wave field, finite length of the shear layer, and other factors. Some results of such investigations that allow for evaluation of the validity of Miles–Ribner solution for plane waves will be analysed in Sect. 4.2 for the sources bounded in space and time near a vortex sheet.

4.2 Excitation of vortex sheet oscillations by an incident wave

Plane monochromatic waves were considered under the assumption that the sound and the normal oscillations of the vortex sheet are linearly independent modes. While acoustic waves from a real source possessing finite size and duration initiate Kelvin–Helmholtz instability on a vortex sheet. Solution of such problems faces both technical difficulties and ambiguity in physical interpretation of results. Already the simplest formulation of the problem of plane pulse reflection from the interface between equilibrium stationary media encounters difficulties because precursors that contradict causality considerations formally appear in the reflected field.

The problem of removing precursors was investigated in ample detail by Kurin *et al.* (1985) who showed that the precursors existing at a given point arbitrarily long as compared to the arrival at this point of the primary mode are formed only when the intersection of the incident wave front and the interface occurs for an infinitely long time. Still greater difficulties emerge in the solution of problems of wave reflection from nonequilibrium media an example of which is a vortex sheet; the problem here is to choose the wave field structure in a transmitted wave, to describe the shape of the reflected signal, and so on.

The difficulties enumerated above can be removed by considering a solution of initial value problem, instead of a stationary solution, employing the Laplace method (Jones and Morgan, 1972; Mironov, 1975).

Let us consider as an example radiation of a monochromatic point source of mass of unit efficiency at a distance h from the vortex sheet (Mironov, 1975). The wave equations in motionless and moving media, respectively, have the forms

$$\begin{aligned} &\Delta\varphi_1 + \left(\frac{\omega}{c_s}\right)^2 \varphi_1 = \delta(x)\delta(y+h), \\ &\Delta\varphi_2 - \frac{1}{c_s^2}\left(-i\omega + U\frac{\partial}{\partial x}\right)^2 \varphi_2 = 0. \end{aligned} \tag{4.11}$$

Taking into account the boundary conditions on the vortex sheet and using the Fourier transform along the x-coordinate one can easily obtain a solution:

$$\varphi = \varphi_0 + \frac{1}{4\pi i}\int \frac{R(\omega,k)}{q_2}\exp[ikx - iq_2(y-h)]\,dk, \tag{4.12}$$

where φ_0 is the source field in an unbounded medium at rest and the reflection coefficient $R(\omega, k)$ is determined by the expression (4.2).

The integration contour in the plane of complex variable k must be chosen proceeding from the causality principle. Since the initial value problem will be solved by the Laplace method, the complex values of ω must correspond to growing waves, i. e., they must lie sufficiently far in the upper half-plane of ω. The poles of the reflection coefficient $R(\omega, k)$ lie in the upper half-plane of complex k. Then, integration may be performed along the real axis of k. For analytical continuation of the solutions to real ω the integration path in the complex plane of k must be deformed by supplementing the real axis with the loops going around the poles k_i in the lower half-plane and with the arcs going around the poles k_s on the real axis (Fig. 4.3).

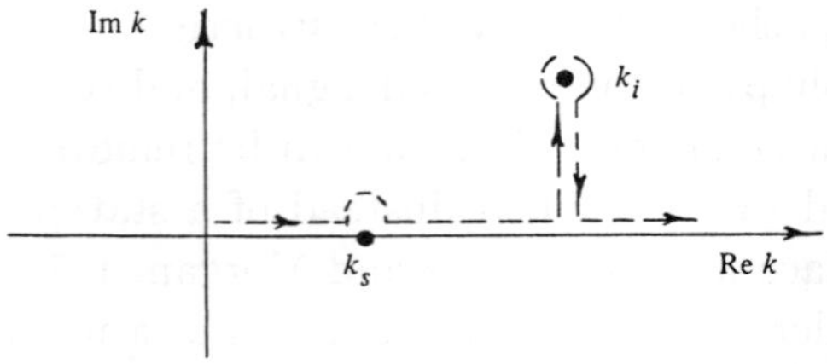

Figure 4.3: The integration contour in the plane of complex k.

The poles k_i correspond to intrinsic oscillations of the vortex sheet which grow along the x-axis. Thus, complete solution of the problem of a point source includes not only progressive sound waves (that are obtained by integration along the real axis in (4.12)) but also the surface wave growing along the x-axis on the vortex sheet.

The solution found by means of this approach enables one to determine the applicability limits of the results related to reflection of plane monochromatic waves from an unstable vortex sheet. Indeed, with the growing distance from the point source ($h \to \infty$), the incident cylindrical wave tends to a plane wave near the direction specified and the efficiency of surface wave excitation decreases exponentially. Note also that isolines of the surface wave amplitude are rays having an angle of incidence $\tan\theta_c = \mathrm{Im}\, q/\mathrm{Im}\, k = 1$ ($\theta_c = 45^o$) (Mironov, 1975). The solution in the form of refracted and reflected waves is meaningless at the angles of incidence $\theta > \theta_c$ because it exists against the background of exponentially growing solution in the form of the surface wave. At the same time, the surface wave may be neglected for $\theta < \theta_c$.

Interesting peculiarities occur when a vortex sheet is excited by a nonstationary source. The source model in the form of an instantaneous point pulse $\delta(t)\,\delta(r)$ that might appear to be very simple at first sight does not hold in the solution of the initial value problem by the Laplace method. Indeed, because the instability rate of normal oscillations of the vortex sheet grows without restriction as the wave number increases, the singular points in the integrand vanish to the infinity along the imaginary axis in the complex plane ω. Therefore, the contour in the plane ω cannot be chosen so that it should pass above all singularities.

The reflection of sound pulses from a vortex sheet was investigated by Nemtsov (1986) for the source of the form $\delta(t)\,\delta(y)\,f(x)$, where $f(x)$ is a bell-shaped function having the spectrum f_k decaying as $k \to \infty$. It was shown that for a two-dimensional stable vortex sheet ($\mathrm{M} \geq 2\sqrt{2}$) the acoustic field is a superposition of a reflected and a lateral wave as well as of three normal modes radiated at the angles θ_0, θ_1, and θ_2.

For an unstable vortex sheet, the incident acoustic pulse excites, besides the lateral and reflected waves, a region of growing perturbations broadening downstream. The instability region in space has a form of a rectangle whose shape depends strongly on the asymptotic form of the spatial source spectrum f_k in the limit $k \to \infty$. This phenomenon referred to as the configuration instability of a vortex sheet was considered in Sect. 3.1.5. The field amplitude inside the instability region grows extremely fast (faster than exponentially) at a rather slow decrease of f_k.

The presence of configuration instability imposes definite restrictions on the use of the vortex sheet model within which the internal structure of the transition region of shear layer between two uniform flows is neglected. The results obtained by means of the vortex sheet model must be used with caution whenever the behaviour of the spectrum in the short-wave region as $k \to \infty$ is concerned with.

4.3 Negative dissipation in a moving medium

The effect of over-reflection, i. e., amplification of a reflected wave by a vortex sheet arises as a result of energy withdrawal from the moving medium thanks to radiation into this medium of NEWs by an oscillating vortex sheet. Over-reflection does not occur when NEWs cannot escape to the moving medium, for example, in a nontransmitting region for $\mathrm{Re}\, q_2 = 0$ or in the presence of a rigid reflecting wall behind the vortex sheet. There exists, however, an alternative situation when over-reflection is realized due to the wave field absorption in a moving medium with dissipation. As was mentioned in Chap. 3, absorption of refracted waves with negative energy occurs in a medium with negative dissipation. It should be born in mind that the dissipation in a moving medium changes its sign together with the wave energy for $\omega - kU < 0$ (see Sect. 1.8). Now consider this effect on an example of viscous dissipation of waves in a moving medium.

The vortex sheets in a viscous medium are eventually replaced by shear layers of finite width where small perturbations have a complex structure (Betchov and Criminale, 1967). However, the effect of viscous dissipation on sound reflection from a moving medium may be evaluated within a vortex sheet model (Kikina, 1967; Miles, 1954). For this purpose consider a vortex sheet between viscous and ideal liquids, i. e., take viscosity into account only in a moving layer when $y > 0$.

Velocity perturbations in a moving viscous medium may be represented in the form

$$u = \frac{\partial \varphi}{\partial x} - \frac{\partial \psi}{\partial y}, \qquad v = \frac{\partial \varphi}{\partial y} + \frac{\partial \psi}{\partial x}. \tag{4.13}$$

The amplitudes of the potential φ and of the stream function ψ meet the following equations (Miles, 1954):

$$\frac{d^2 \varphi}{dy^2} + q_+^2 \varphi = 0, \qquad \frac{d^2 \varphi}{dy^2} + q_\nu^2 \varphi = 0, \tag{4.14}$$

where

$$q_+ = \sqrt{\frac{(\omega - kU)^2}{1 - 4i\nu(\omega - kU)/3c_s^2} - k^2}, \qquad q_\nu = \sqrt{\frac{i(\omega - kU)}{\nu} - k^2},$$

and the pressure amplitude is

$$p = \frac{i(\omega - kU)\rho\varphi}{1 - 4i\nu(\omega - kU)/3c_s^2}. \tag{4.15}$$

Thus, the general solution

$$u = ik\varphi_0 e^{iq_+ y} - iq_\nu \psi_0 e^{iq_\nu y},$$
$$v = iq_\nu \varphi_0 e^{iq_+ y} + ik\psi_0 e^{iq_\nu y} \tag{4.16}$$

is a superposition of two modes: a potential one which is an acoustic wave modified by viscosity and a vortex mode for which there are no pressure oscillations in a linear approximation. At large Reynolds numbers, $\mathrm{Re} = \nu|\omega - kU|/c_s^2$, a vortex viscous mode oscillates rapidly and damps as $y \to \infty$.

The dynamic boundary condition on a vortex sheet is continuity of the momentum flux components normal to the vortex sheet, i. e., of the components of stress tensor (Landau and Lifshits, 1986):

$$\sigma_{xy} = \rho\nu\left(\frac{\partial u}{\partial y} + \frac{\partial v}{\partial x}\right),$$
$$\sigma_{yy} = -p + 2\rho\nu\frac{\partial v}{\partial y} - \frac{2}{3}\rho\nu\left(\frac{\partial u}{\partial y} + \frac{\partial v}{\partial x}\right). \tag{4.17}$$

These boundary conditions together with the kinematic condition (continuity of the displacement $\eta = iv/(\omega - kU)$ at the intersection of the vortex sheet surface) enable one to match the viscous solutions (4.15) and (4.16) for $y > 0$ and the solutions (4.1), (4.2) for φ_-, p_- in an ideal liquid at rest for $\mathrm{Re}^{-1} \ll 1$. This yields the coefficient of sound reflection in the first approximation with respect to the parameter Re^{-1}:

$$R = \frac{(1 - kU/\omega)^2 q_-/q_+ - 1 + 4i\nu k^2/(\omega - kU)}{(1 - kU/\omega)^2 q_-/q_+ + 1 - 4i\nu k^2/(\omega - kU)}. \tag{4.18}$$

The expression (4.18) transforms to (4.2) as $\nu \to 0$. In the region of total reflection, where the reflection coefficient $|R| = 1$ at $\nu = 0$, the correction due to viscosity determines the value of relative dissipation:

$$1 - |R|^2 = \frac{16\nu k^2(\omega - kU)q_-|q_+|^2\omega}{q_-^2(\omega - kU)^4 + |q_+|^2\omega^4}. \tag{4.19}$$

The dissipation in a moving medium becomes negative when $\omega < kU$, which leads to amplification of the reflected wave: $|R| > 1$.

4.4 Over-reflection of internal waves

Generation of IGWs by shear flows in a stratified medium is, apparently, an example of the wave–hydrodynamic flow interaction that is most significant for geophysical applications because the phase velocity of IGWs is frequently comparable with the velocity of real flows in an atmosphere and in the ocean. Consider here the possibility of IGW amplification due to over-reflection on a vortex sheet (McKenzie, 1972; Troitskaya and Fabrikant, 1987).

Equations for two-dimensional oscillations of a stratified incompressible medium moving with a velocity U can be written in the form (Gossard and Hooke, 1975):

$$\begin{aligned} \frac{\partial u}{\partial t} + U\frac{\partial u}{\partial x} + \frac{1}{\rho}\frac{\partial \rho}{\partial x} &= 0, \\ \frac{\partial v}{\partial t} + U\frac{\partial u}{\partial x} + \frac{1}{\rho}\frac{\partial \rho}{\partial y} + \frac{g\tilde{\rho}}{\rho} &= 0, \\ \frac{\partial \tilde{\rho}}{\partial t} + U\frac{\partial \rho}{\partial x} + v\frac{\partial \rho}{\partial y} &= 0, \\ \frac{\partial u}{\partial x} + \frac{\partial v}{\partial y} &= 0. \end{aligned} \tag{4.20}$$

For IGWs the Boussinesq approximation corresponding to the limit

$$\frac{d\rho}{dy} \to 0, \quad g \to \infty \quad \text{with} \quad N^2 \equiv \frac{-g}{\rho}\frac{d\rho}{dy} = \text{const}$$

is usually valid. Within this approximation, density variation along the wavelength scale is insignificant but the restoring buoyancy force is conserved. The derivative $d\rho/dy$ enters only the Brunt–Väisälä frequency N and ρ may be assumed to be constant in all the coefficients of the equations. Equations for the amplitudes of the two-dimensional small perturbations proportional to $\exp(-i\omega t + ikx)$ in a fluid moving with a velocity $U = (U, 0, 0) =$ const have, in the Boussinesq approximation, the following form

$$\begin{aligned} -i(\omega - kU)u + ikp &= 0, \\ -i(\omega - kU)v\left[1 - \frac{N^2}{(\omega - kU)^2}\right] + \frac{\partial p}{\partial y} &= 0, \qquad (4.21) \\ iku + \frac{\partial v}{\partial y} &= 0. \end{aligned}$$

From (4.21) one can obtain a dispersion equation for plane IGWs with the amplitude proportional to $\exp(iqy)$ and an equation for their group velocity c_g:

$$(\omega - \mathbf{k}\mathbf{U})^2 = \frac{N^2 k^2}{k^2 + q^2} = N^2 \sin^2\theta,$$

$$\mathbf{c}_g = \frac{Nq}{(k^2 + q^2)^{3/2}}(q_1, -k) = \frac{N\cos\theta}{k_0}(\cos\theta, -\sin\theta), \qquad (4.22)$$

$$\mathbf{k} = k_0(\sin\theta, \cos\theta), \qquad (\mathbf{k}, \mathbf{c}_g) = 0.$$

Note that uniform plane IGWs may propagate only for $\omega - kU < N$.

Consider now the reflection of IGWs from a vortex sheet separating media of different stratifications:

$$U(y) = \begin{cases} U_+ & \text{at} \quad y > 0; \\ U_- & \text{at} \quad y \le 0; \end{cases} \qquad N(y) = \begin{cases} N_+ & \text{at} \quad y > 0; \\ N_- & \text{at} \quad y \le 0. \end{cases} \qquad (4.23)$$

The formulation of this problem is fully analogous to the Miles–Ribner problem of sound reflection. The components of velocity perturbations may be expressed through the stream function ψ:

$$u = \partial\psi/\partial z, \qquad v = -ik\psi, \qquad (4.24)$$

which in the upper and lower half-planes has the following form

$$\begin{aligned} \psi(y > 0) &= Te^{iq_+ z}, \\ \psi(y < 0) &= e^{iq_- z} + Re^{-iq_- z}, \end{aligned} \qquad (4.25)$$

where $q_{\pm}^2 = k^2 \left[N_{\pm}^2/(\omega - kU_{\pm})^2 - 1\right]$. The values of the reflection coefficient R and the transmission coefficient T may be found, like in the Miles–Ribner problem, by matching the solutions (4.25) by means of the boundary conditions of pressure and displacement continuity: $p_+ = p_-$, $\zeta_+ = \zeta_-$ on the surface of the vortex sheet $z = 0$, which give

$$\begin{aligned} R &= \frac{q_-(U_- - \omega/k)^2 - q_+(U_+ - \omega/k)^2}{q_-(U_- - \omega/k)^2 + q_+(U_+ - \omega/k)^2}, \\ T &= \frac{U_+ - \omega/k}{U_- - \omega/k}\frac{2q_-(U_- - \omega/k)^2}{q_-(U_- - \omega/k)^2 + q_+(U_+ - \omega/k)^2}. \end{aligned} \tag{4.26}$$

The law of energy conservation

$$\frac{q_-\left(1 - |R|^2\right)}{N_-^2 - (\omega - kU_-)^2} = \frac{q_+|T|^2}{N_+^2 - (\omega - kU_+)^2} \tag{4.27}$$

that follows from (4.26) can easily be obtained from the expression (4.22) for c_g and from the formula for the energy density of IGWs in a medium at rest (Gossard and Hooke, 1975):

$$E_0 = \frac{\rho\langle u^2 + v^2\rangle}{2} + \frac{N^2\langle\eta^2\rangle}{2} = \frac{N^2\,k^2}{\omega^2(N^2 - \omega^2)}|p|^2, \tag{4.28}$$

with (2.4) and the relation $S_y = (c_g)_y E$ taken into account.

Let us assume further that the wave is incident from a medium at rest $U_- = 0$. The values of vertical components of the wave vector $q_{\pm}$ determine different regimes of reflection. The vertical component of the group velocity of IGWs in an uniformly moving medium has the following form

$$(c_g)_y = -\frac{N^2\,k^2\,q}{(\omega - \mathbf{kU})(k^2 + q^2)^2}. \tag{4.29}$$

The radiation condition $(c_g)_y > 0$ allows one to choose the branch of the function $q(\omega, k)$. This condition in (4.25) demands to choose $q_- < 0$ for a wave in a motionless medium. In a moving medium $(y \to +\infty)$, the incident wave is reflected completely ($\operatorname{Re} q_+ = 0$) at $(\omega - kU_+)^2 > N_+^2$ while at $(\omega - kU_+)^2 < N_+^2$ the radiation condition $(c_g^+)_y > 0$ determines the sign of q_+. When $\omega - kU_+ < 0$ one has $q_+ > 0$. In this case, as follows from (1.4), the refracted wave has negative energy, which gives rise to IGW over-reflection from a vortex sheet ($|R| > 1$).

Different regimes of IGW reflection from a vortex sheet may be classified as follows. The behaviour of the transmitted field is determined by two dimensionless parameters: the quantity $D = N_+/N_-$ that characterizes the jump in stratification and the dimensionless phase velocity $\tilde{c} = \omega/kU_+ = N_-/k_0U_+$. Figure 4.4 presents partitioning of the plane of the parameters $(D, 1/\tilde{c})$ into regions for which the regimes of IGW refraction are shown in Fig. 4.5 depending on the angle of incidence θ.

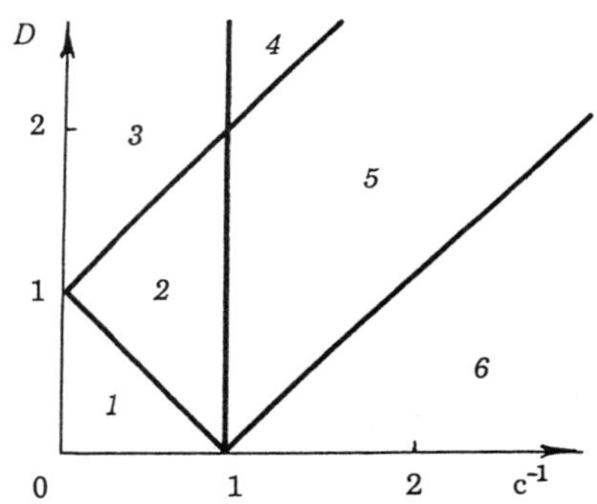

Figure 4.4: Partitioning of the plane of vortex sheet parameters into regions with different regimes of IGW reflection by a vortex sheet.

The critical angles $\tilde{\theta}_{1,2}$ bounding the regions of total internal reflection at which the transmitted wave is absent are determined by the following expression

$$\sin\tilde{\theta}_1 = \frac{D\tilde{c}}{\tilde{c}-1}, \qquad \sin\tilde{\theta}_2 = -\frac{D\tilde{c}}{\tilde{c}+1}. \tag{4.30}$$

Over-reflection regimes are realized only at sufficiently short waves when IGWs satisfy the condition $k_0 > N_-/N_+$.

Of particular interest are vortex sheet resonances ($|R| = \infty$) that can easily be found in the case of uniform stratification: $N_1 = N_2$. They are possible only for $\tilde{c} > 1$. One of them is determined by the condition

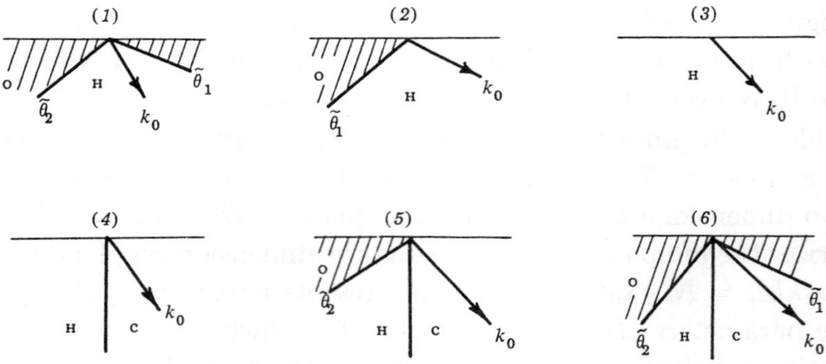

Figure 4.5: Regimes of IGW reflection. The number of the graph corresponds to the number of the region on the plane of the parameters D, $1/\tilde{c}$. The ranges of the angles of incidence determined by the wave vector of the incident wave are marked as in Fig. 4.2.

$\tilde{c} = 1/2$ $(\omega = kU/2)$ and corresponds to "Cherenkov" radiation of IGW by a vortex sheet. The angle of radiation is then determined by the relation $\sin\theta = \omega/N_1 = kU/2N_1$. Two other resonances are determined by the condition $\cot\theta = 1/\tilde{c} - 1$, which corresponds to the dispersion equation

$$\omega^2 + (\omega - kU)^2 = N_1$$

for normal waves on a vortex sheet. These modes are stable for wave numbers $|k| < \sqrt{2}\,N_1/U$ and are transformed to the Kelvin–Helmholtz modes as $N_1 \to 0$.

4.5 Over-reflection of surface gravity waves

Waves on water surface is another example that is highly significant for geophysics. Possible mechanisms of generation of surface gravity waves (SGW) due to their interaction with air flow or with shear surface flow in water were considered in Sect. 2.3. In this section we will consider still another mechanism of interaction of surface waves with a hydrodynamic flow — over-reflection of surface waves incident on a vortex sheet — that may be interesting for the physics of the ocean.

A shallow water approximation (Landau and Lifshits, 1986) is valid for long waves ($kH \ll 1$). The problem of surface wave refraction and reflection is equivalent, in this case, to the acoustic Miles–Ribner problem. Note, however, that in contrast to acoustics, the propagation of surface waves on shallow water is essentially two-dimensional. Correspondingly, the remarks made in Sect. 4.1 concerning the three-dimensional disturbances destabilizing a supersonic vortex sheet do not refer to the problem considered here. This is exactly the case when the conclusion made by L. D. Landau about the stability of a supersonic vortex sheet is valid: a vortex sheet on a shallow water becomes stable with respect to two-dimensional perturbations when (Nezlin and Snezhkin, 1993)

$$\mathrm{M} \equiv \frac{U}{\sqrt{gH}} \geq 2\sqrt{2}.$$

The problem of wave reflection on deep water proves to be much more complicated. The corresponding boundary value problem consists in the solution of a system of integral equations which were first derived by Evans (1975). Solution of this problem involves very intricate mathematics. An approximate solution using Galerkin's method was found by Basovich (1978). Employing the approximation proposed by Miles (1967) he found an expression for the reflection coefficient of the sinusoidal ($\sim e^{-i\omega t+ik\pm iqy}$) surface waves:

$$R = \frac{(q_-/q_+)S^2 - 1 - iX}{(q_-/q_+)S^2 + 1 - iX}, \tag{4.31}$$

where

$$q_- = \frac{\omega^2}{g}\cos\theta_-,$$

$$q_+ = \frac{\omega^2}{g}(1 - \mathrm{M}\sin\theta_-)\sqrt{(1-\mathrm{M}\sin\theta_-)^2 - \frac{\sin^2\theta_-}{(1-\mathrm{M}\sin\theta_-)^2}}$$

are the projections of the wave vectors of incident and refracted waves on the y-axis, S and X are the real functions of the angle of incidence

θ_- and of the angle of refraction θ_+ as well as of the parameter (that is analogous to Mach number in acoustics) relating the velocity jump to the phase velocity of the gravity wave: $M = \omega U/g$.

The asymptotic behaviour of the refracted field as $y \to \infty$ and the regime of reflection are determined by the wave number q_+. Normal reflection: $q_+ > 0$; $|R| < 1$ occurs at

$$\mathsf{M} < \mathsf{M}_1 \equiv \frac{1 - \sqrt{\sin\theta_-}}{\sin\theta_-}.$$

Full reflection: $\mathrm{Re}\, q_+ = 0$; $|R| = 1$ takes place in the range

$$\mathsf{M}_1 < \mathsf{M} < \mathsf{M}_2 \equiv \frac{1 + \sqrt{\sin\theta_-}}{\sin\theta_-}.$$

Finally, for $M > M_2$ the regime of over-reflection appears when $q_+ < 0$ and $|R| > 1$. In the latter case, $\omega - kU_+ = \omega(1 - M\sin\theta_+) < 0$ for the refracted wave and its energy is negative. The phase velocity of the refracted wave is directed towards the vortex sheet while the wave packet escapes from it moving in the direction of the group velocity whose projection is

$$(\mathbf{c}_g^+)_y = \frac{q_+ g^2}{2(\omega - kU)^3} > 0.$$

For $\sin\theta_- = 2/M$ (i. e., at $|\omega - k\,U| = \omega$), the incident wave resonates with the normal mode of the vortex sheet and $|R| = \infty$. The only normal mode of frequency ω exists for $M > 2$ and is a nonuniform plane wave with a wave vector directed at an angle of $\gamma = \pi/2 - \theta_- = \arccos(2/M)$ to the vortex sheet. The group velocity of this wave has different directions on either side of the vortex sheet. This mode is essentially the result of "Cherenkov" radiation of perturbations on a vortex sheet. It is interesting to note that the normal mode of the vortex sheet found by Basovich (1978) is, actually, an exact solution that has no relation to the approximation used by this author.

It should be noted in conclusion that over-reflection of surface waves may give rise to the instability of jet flows relative to perturbations of a free surface.

4.6 Rossby waves on a vortex sheet

The presence of a hydrodynamic flow with sufficiently great velocity that is able to "overtake" the waves incident on it does not yet guarantee over-reflection which, besides, needs radiation into a moving medium of a refracted wave of negative energy. Consider as an illustration Rossby waves incident on a vortex sheet on the β-plane (Fabrikant, 1987).

The perturbation of the stream function ψ in an incompressible rotating medium on the β-plane obeys the following equation

$$\frac{\partial}{\partial t}\Delta\psi + \beta\frac{\partial\psi}{\partial x} = 0. \tag{4.32}$$

This equation holds for the perturbations whose characteristic frequencies are small compared to the frequency with which the medium rotates: $\Omega(y) = \Omega_0 + \beta y$ (a quasi-geostrophic approximation). We would like to draw your attention to the analogy with drift waves in an inhomogeneous plasma which, under certain conditions, are described by equations in a geostrophic approximation (Hasegawa and Mima, 1978). Consequently, the results that will be presented below are also valid for drift waves on a vortex sheet.

From (4.32) follow a dispersion equation and a group velocity for Rossby waves:

$$\begin{aligned} \omega &= -\frac{\beta k}{k^2+q^2}, \\ \mathbf{c}_{\mathrm{g}} &= \frac{\beta}{(k^2+q^2)^2}(k^2-q^2, 2kq), \end{aligned} \tag{4.33}$$

where k and q are the x, y-components of the wave vector $\mathbf{k} = (k, q)$, respectively.

Only waves possessing negative phase velocity along the x-axis and having the wave number $k > 0$ are possible for $\beta > 0$. The y-components of the phase and group velocities have opposite signs:

$$q(c_{\mathrm{g}})_{\mathrm{y}} = \frac{2\beta q^2}{(k^2+q^2)^2}k < 0.$$

These relations determine the wave vector directions in the incident, reflected and refracted waves. Thus, assuming the liquid to be at rest for $y < 0$, one can consider the wave field in the form (4.25) where the y-coordinate, instead of the vertical z-coordinate, must be taken as an independent variable and the y-components of the wave vector have the form

$$q_1 = -\sqrt{-\frac{\beta k}{\omega} - k^2} < 0, \qquad q_2 = -\sqrt{-\frac{\beta k}{\omega - kU} - k^2} < 0.$$

The boundary conditions on the vortex sheet — continuity of two values:

$$\left[\frac{\psi}{\omega - kU}\right]_-^+ = 0, \qquad \left[(\omega - kU)\frac{d\psi}{dy}\right]_-^+ = 0, \tag{4.34}$$

allow for matching the solutions (4.25) at the boundary $y = 0$, which yields

$$\begin{aligned} R &= -\frac{q_-\omega^2 - q_+(\omega - kU)^2}{q_-\omega^2 + q_+(\omega - kU)^2}, \\ T &= \frac{2q_-\omega(\omega - kU)}{q_-\omega^2 + q_+(\omega - kU)^2}. \end{aligned} \tag{4.35}$$

From the conditions $q_{1,2} < 0$ follows that $|R|$ is always smaller than unity, i. e., no over-reflection occurs in this case. This is explained by the fact that the energy density of the refracted wave propagating in a moving medium is positive. Actually, we have

$$\begin{aligned} E &= \frac{\omega}{\omega - kU}E_0 \\ &= \frac{\omega}{\omega - kU}(k^2 + q_+^2)|\psi|^2 = \frac{\omega(k^2 + q_+^2)^2}{-\beta k}|\psi|^2 > 0. \end{aligned} \tag{4.36}$$

Various regimes of reflection for $\beta > 0$ are shown in Fig. 4.6 depending on the parameter $\tilde{c} = \omega/kU_+$. The regions of normal and total reflection are separated by the angle $\theta = \arcsin\sqrt{\tilde{c}/(\tilde{c} - 1)}$.

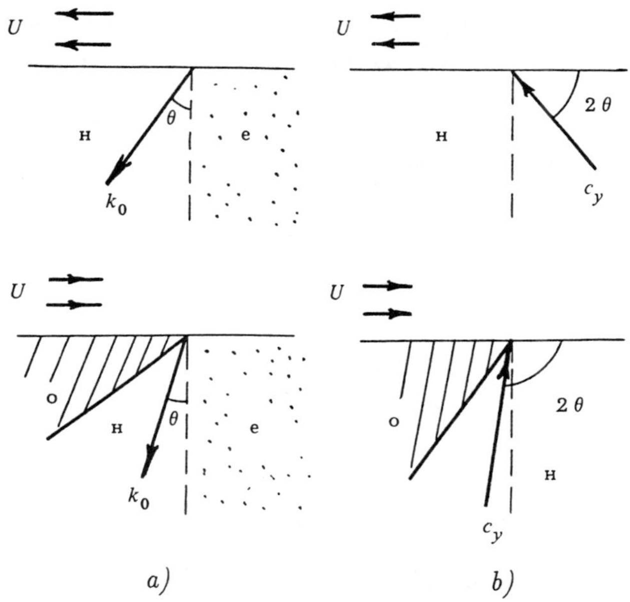

Figure 4.6: Regimes of the Rossby wave reflection at a vortex sheet. Marked are the ranges of the angles of incidence for the wave vector (a) and group velocity (b): e – no propagation, o – total reflection, H – normal reflection.

The condition $|R| = \infty$ gives a dispersion equation for normal waves on a vortex sheet in a rotating medium:

$$(\omega - kU)^2\sqrt{k^2 + \frac{\beta k}{\omega - kU}} + \omega^2\sqrt{k^2 + \frac{\sqrt{\beta k}}{\omega}} = 0. \qquad (4.37)$$

Designating $c_0 = 2\omega/kU - 1$, one can readily obtain from (4.37) a cubic equation for this dimensionless parameter:

$$c_0^3 + 3sc_0^2 + c_0 + s = 0, \text{ where } s = \frac{\beta}{2k^2U}.$$

The discriminant of this equation is $\Delta = -108\,s^4 + 36\,s^2 - 4 < 0$, consequently it has two complex–conjugate roots. One of these roots having positive imaginary part is responsible for the instability. The plots for Re c_0 and Im c_0 versus parameter s are depicted in (Fig. 4.7).

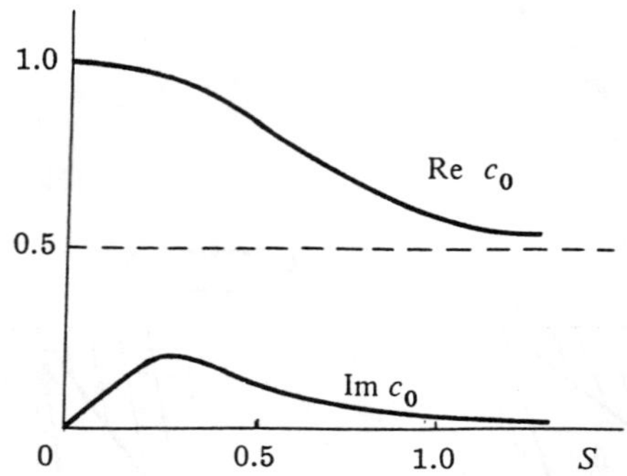

Figure 4.7: Complex phase velocity of unstable perturbations on a vortex sheet in a rotating medium.

Thus, vortex sheet instability is conserved in the presence of medium rotation and transforms to an ordinary Kelvin–Helmholtz instability as $s \to 0$.

Chapter 5

Resonant Wave-Flow Interaction

Wave propagation in shear flows is often accompanied by the effects associated with their amplification and absorption caused by resonances, i. e., the layers in which the flow velocity coincides with the phase and/or group velocities of the waves. These are, in the first place, critical layers which play a key role in the mechanism of hydrodynamic instability of shear flows (see Chap. 2).

Resonant interaction of waves and critical layers was considered as a significant mechanism of generation and absorption of different types of waves propagating in the atmosphere (Gossard and Hooke, 1975; Holton, 1975; Ahmed and Eltaeb, 1988), in the ocean (Phillips, 1977), in a moving plasma (Timofeev, 1971, 1979; Gavrilenko and Zelekson, 1977), etc.

A progressive wave may also strongly interact with a flow in the layers where the wave frequency coincides with the frequencies of normal oscillations of an inhomogeneous flow. In this case, the resonance is caused by the equality of the group velocity of the wave and the velocity of a definite layer of liquid, which gives rise to wave beam trapping in the neighborhood of the resonance.

The algebraic method considered in Chap. 3 for the investigation of waves in shear flows by means of a piece-wise linear approximation of velocity profile does not hold for the analysis of resonance effects because it neglects the presence of resonance singularities in governing equations.

Consequently, the waves in a flow with a continuous velocity profile are investigated by means of numerical (Betchov and Criminale, 1967) or approximate asymptotic methods: the WKB method and the method of reference equations (Gavrilenko *et al.*, 1975; Basovich and Tsimring, 1984), as well as the method of matched asymptotic expansions (Andronov and Fabrikant, 1980). Some results may also be obtained employing model profiles (Jones, 1968; Fabrikant, 1976; Erokhin and Sagdeev, 1985). Finally, the plasma–hydrodynamic analogy may also be very fruitful, since it is a handy tool for qualitative physical understanding of resonance processes.

5.1 Resonant sound amplification

Investigation of sound reflection at a plane-parallel flow having a continuous velocity profile is a much more difficult task than the Miles–Ribner problem of reflection at a vortex sheet (Steinmets and Singh, 1972; Shubert, 1972; Fabrikant, 1976; Gavrilenko and Zelekson, 1977; Andronov and Fabrikant, 1980). Consider an unperturbed shear flow with a velocity profile $U(y)$, for which the velocity tends to zero, as $y \to -\infty$, in the lower half-plane and to a constant, as $y \to \infty$, in the upper half-plane. The behavior of small perturbations in the region of displacement ($|y| \leq 1$) is determined by Rayleigh's equation for pressure amplitude:

$$\frac{d^2p}{dy^2} - \frac{2}{U-c}\frac{dU}{dy}\frac{dp}{dy} + k^2\left[\frac{(U-c)^2}{c_s^2} - 1\right]p = 0. \qquad (5.1)$$

Assume that the sources of perturbations are available only for $y \to -\infty$. Then, the expressions (4.1) are the asymptotic forms of the solutions in the limit $y \to \pm\infty$.

Equation (5.1) for supersonic flows, like Eq. (3.1) for an incompressible fluid, may contain singular points y_0 at which $U(y_0) = c \equiv \omega/k$

and the solution $p(y)$ has, in general, branching points. The choice of a branch of a multivalued solution in the neighborhood of y_0 (critical layer) is determined by Lin's rule of integration path.

The resonant interaction of a sound wave with a flux that occurs in a critical layer as $y \to y_0$ gives rise to the absorption or amplification of sound energy. A criterion indicating the direction of energy exchange between the wave and the flow may be obtained from (5.1) in a general form. Multiplication of (5.1) by $p^*/(U-c)^2$ (where p^* is a complex conjugate function) and integration by parts, with Lin's rule and the boundary conditions (5.1) taken into account, yield the law of energy conservation in the form

$$\frac{q_-}{c^2}\left(1-|R|^2\right) - \frac{q_+}{(U-c)^2}|T|^2 = \pi k^2 \left[|p|^2 \frac{d^2U/dy^2}{|dU/dy|^3}\right]_{y=y_c}. \tag{5.2}$$

The left-hand side of this equation describes relative absorption of sound and represents the difference between the energy flux of an incident wave (with unit amplitude) and the sum of energy fluxes for reflected and refracted waves. The right-hand side of (5.2) describes the energy absorption in a critical layer. The value of absorption in this case, as for wind waves (cf. Sect. 2.2), is proportional to the derivative of the velocity distribution function of the resonance particles of a shear flow, i. e., to the curvature of the velocity profile at the resonance point. The concave portions of the profile ($d^2U/dy^2 < 0$) amplify the sound wave while the convex portions ($d^2U/dy^2 > 0$) absorb it. The position of the critical layer at a given velocity profile of sound waves depends on the angle of incidence:

$$\sin\theta = c_s/c = c_s/U(y_c).$$

This position specifies uniquely the sign of resonant interaction as well.

The reflection and refraction coefficients of sound for an arbitrary continuous profile may be calculated analytically, given that a small parameter is available in the boundary value problem (5.1). In the case of a "narrow" shear layer when the characteristic length of the flow, l, is small compared to the wavelength, $\lambda = 2\pi/k$, i. e., when $\mu = kl \ll 1$, the solutions to Eq. (5.1) may be found employing the method of matched asymptotic expansions (Van Dyke, 1975).

Consider a shear layer having a monotonic velocity profile $U(y/l)$ with the asymptotic form exponentially tending to constant values as $|y| \to \infty$:

$$\begin{gathered} |U(y) - U(\pm\infty)| \sim \exp(-2|y|/l), \\ U(-\infty) = 0, \quad U(+\infty) = U_0, \quad dU/dy > 0. \end{gathered} \tag{5.3}$$

Let us introduce internal, ξ, and external, σ, variables:

$$\xi = y/l, \qquad \sigma = ky = \mu\xi. \tag{5.4}$$

The subscripts i and e will designate internal and external expansions and the figures, the order of these expansions in the small parameter μ. The external asymptotic series expansion in this parameter may be found from Eq. (5.1) written in the dimensionless form

$$\frac{d^2p}{d\sigma^2} - \left[1 - \frac{(U-c)^2}{c_s^2}\right]p = \frac{2}{U(\sigma/\mu) - c}\frac{dU(\sigma/\mu)}{d\sigma}\frac{dp}{d\sigma}. \tag{5.5}$$

The right-hand side of this equation is exponentially small (proportional to $e^{-2|\sigma|/\mu}$) for $\sigma \geq 1$. Therefore, it does not contribute to the asymptotic power series in μ. Taking into account the boundary conditions, as $y \to \pm\infty$, one finds the first-order approximationwith respect to μ:

$$\begin{aligned} p_{e1}^+ &\equiv p_{e1}(\sigma > 0) = T\exp(iq_2\sigma/k), \\ p_{e1}^- &\equiv p_{e1}(\sigma < 0) = \exp(iq_1\sigma/k) + R\exp(-iq_1\sigma/k). \end{aligned} \tag{5.6}$$

Let us now express the found solution through the internal variable ξ and take the second order-approximation with respect to μ:

$$\begin{aligned} (p_{e1}^+)_{i2} &= T - i\mu\xi T q_2/k - \mu^2 T\xi^2 q_2^2/2k^2, \\ (p_{e1}^-)_{i2} &= 1 + R - i\mu(1-R)\xi q_1/k - \mu^2(1+R)\xi^2 q_1^2/2k^2. \end{aligned} \tag{5.7}$$

Let us now seek the second-order internal expansion using in (5.1) the internal variable ξ:

$$\frac{d^2p}{d\xi^2} - \frac{2}{U-c}\frac{dU}{d\xi}\frac{dp}{d\xi} = \mu^2\left[1 - \frac{(U-c)^2}{c_s^2}\right]p. \tag{5.8}$$

Assuming that the right-hand side of this equation is equal to zero one can find its formal solution:

$$p = E + \mu F\int (U-c)^2\, d\xi. \tag{5.9}$$

The second component in (5.9) is proportional to ξ when $\xi \gg 1$. Consequently, for subsequent matching of (5.9) to the external solution (5.7) this component is taken with the coefficient μ. The second-order approximation with respect to μ may be found by means of successive approximations by solving (5.8). The resulting internal second-order expansion is written in the form:

$$\begin{aligned} p_{i2} &= E + \mu F\int_0^{\xi}(U-c)^2\, d\xi \\ &+ \mu^2 E\left\{\int_0^{\xi}[U(\tau)-c]^2\int_0^{\tau}\frac{d\xi d\tau}{(U-c)^2} - \int_0^{\xi}\frac{(U-c)^2}{c^2}\tau\, d\tau\right\}. \end{aligned} \tag{5.10}$$

On performing some tedious and awkward but trivial enough transformations one can express this internal expansion through the external variable σ and conserve the first-order approximation with respect to μ:

$$\begin{aligned} (p_{i2})_{e1}^{+} &= E + \mu F\left[W_2^{+} + \xi(U_0-c)^2\right] + \mu^2 E\left[\xi(U_0-c)^2W_{-2}^{+} - \frac{\xi^2q_2^2}{2k^2}\right], \\ (p_{i2})_{e1}^{-} &= E - \mu F\left[W_2^{-} - \xi c^2\right] - \mu^2 E\left[\xi c^2W_{-2}^{-} + \frac{\xi^2q_1^2}{2k^2}\right], \end{aligned} \tag{5.11}$$

where for $i = 1, 2$ one has

$$W_i^{\pm} = \pm \int_0^{\pm\infty} \left\{ (U - c)^i - [U(\pm\infty) - c]^i \right\} d\xi.$$

Let us now "match" the internal and external asymptotic expansions, i. e., let us equate (5.7) and (5.11). The expression for the reflection coefficient of the sound wave in the first-order approximation with respect to μ is then written in the form

$$R = \frac{q_1/\omega^2 - q_2 S/(U_0 - c)^2}{q_1/\omega^2 - q_2 S/(U_0 - c)^2}, \tag{5.12}$$

where

$$S = 1 - i\mu \frac{k}{q_2}(U_0 - c)^2 W_{-2} - i\mu \frac{q_2}{k}(U_0 - c)^2 W_{-2}; \quad W_i = W^+ i + W_i^-.$$

The expressions of the form (5.12) for the coefficients R and T differ from the corresponding formulas (5.4) for a vortex sheet by a small value μ. These corrections are, actually, due to the effect of finite thickness of the velocity profile on sound wave reflection and are described by the integrals throughout the region of velocity shift. The integrals $W_i^{\pm}$ contain, in general, the poles in the integrand which correspond to the resonance points y_c at which $U(y_c) = c$. According to Lin's rule, the integration path around the poles in the complex plane determines the imaginary part of these integrals:

$$\mathrm{Im}\,(\mu W_{-2}) = \mathrm{Im}\left[k \int \frac{dy}{(U - c)^2}\right] = k \sum \left[\frac{U''(y)}{|U'(y)|^3}\right]_{y=y_c}, \tag{5.13}$$

where the sum is taken over all the resonance points y_c. The quantity (5.13) describes the wave amplification on reflection from a shear layer. In particular, $|R| = 1$ for the vortex sheet in the region of total reflection at the moving layer in the absence of a refracted wave. While now the value of relative amplification of the reflected wave

$$1 - |R|^2 \sim \mu\,\mathrm{Im}\,(W_{-2}) \tag{5.14}$$

depends solely on the resonant interaction with synchronous particles. The mechanism of over-reflection associated with the presence of reflected NEWs does not work in this case.

Another limiting case of smoothly inhomogeneous flows ($\mu \gg 1$) also allows for analytical solution. The value of resonance interaction is small in this case because the critical level of y_c lies in the nontransparent region where the wave field is decaying exponentially in the penetration deep inside the region.

Evidently, the resonance effect reaches its maximum at $\mu \gg 1$ and depends on the velocity profile.The reflection and refraction coefficients for the shear layer having the profile

$$U(y) = \frac{U_0}{2}\left(1 + \tanh\frac{y}{l}\right) \tag{5.15}$$

were calculated numerically by Fabrikant (1976). The reflection coefficient as a function of the angle of incidence (in the angular range of total reflection) is plotted in Fig. 5.1 for different values of the parameter μ and the Mach number M = 3.

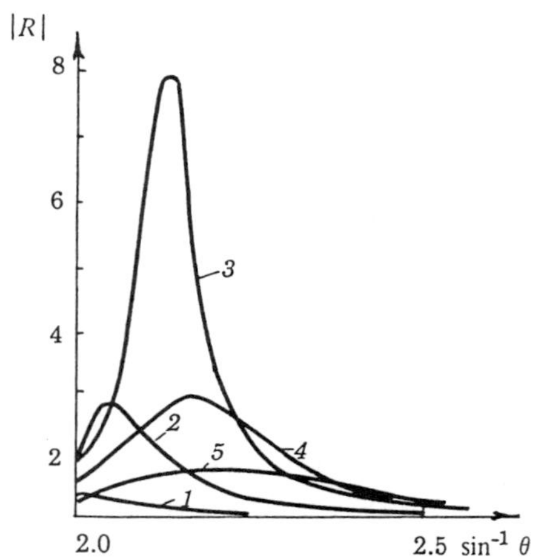

Figure 5.1: Angular dependence of acoustic reflection coefficient:
(1) $\mu = 0.1$, (2) $\mu = 0.3$, (3) $\mu = 0.7$, (4) $\mu = 1.0$, (5) $\mu = 1.5$.

Resonance sound amplification may amount to rather high values.

Thus, for $\mu = 0.7$, the value of relative amplification in the considered angular range is $(|R|^2 - 1) \sim 10^2$.

5.2 IGW amplification in a critical layer

The resonant interaction with a shear flow in a critical layer is a very important mechanism of absorption and generation of atmospheric waves. In particular, resonant absorption affects significantly the dissipation of IGWs propagating from the troposphere, where they are generated by different mechanisms, through the regions of pronounced wind shifts to the middle and upper atmosphere (Gossard and Hooke, 1975). The investigation of resonant absorption of IGWs initiated by Booker and Bretherton (1967) was later developed for different shear flows.

The value of Richardson's number $\mathrm{Ri} = N^2(dU/dy)^{-2}$ in a critical layer plays a fundamental role for IGWs in shear flows. Given $\mathrm{Ri}_c = \mathrm{Ri}(y_c) > 1/4$, the wave energy is absorbed in the critical layer independent of the velocity profile. The interpretation of this result is based on the WKB method in the neighborhood of the critical layer ($\mathrm{Ri}_c \gg 1/4$) where the wave packet of diminishing wavelength is infinitely long approaching the critical point.

At the same time, when $\mathrm{Ri}_c < 1/4$, the resonant interaction of IGW and a flow in a critical layer may give rise to wave amplification. In particular, in the absence of stratification ($N = 0, \quad \mathrm{Ri} = 0$) near the critical layer, the sign of energy dissipation depends, as was shown above, on the curvature of the velocity profile: the waves are amplified at $d^2U(y)/dy^2 < 0$. The authors of the few papers (Jones, 1968; Eltaeb and McKenzie, 1975; Van Duin and Kelder, 1982; Linzen and Barker, 1985) that were concerned with resonant IGW amplification either employed model (piecewise-linear, etc.) velocity profiles and numerical methods or sought for some particular flows. Analytical investigation of resonant IGW amplification was undertaken for flows with smooth velocity profiles by Lin (1955) and Troitskaya and Fabrikant (1989) who also gave the physical interpretation of the mechanism of resonant IGW–flow interaction. Below we will use Lin's approach for the investigation of resonant IGW amplification in a flow with an arbitrary "narrow" velocity profile by means of the method of matched asymptotic expansions, similarly to the case of sound waves.

Consider a two-dimensional problem of reflection and refraction of plane monochromatic IGWs in a stratified shear layer having a velocity profile $U(y)$. This problem is analogous to the problem considered in Sect. 4.1 for a sound wave. The amplitude of the harmonic perturbation of the stream function, ψ, in the Boussinesq approximation satisfies the Taylor–Goldstein equation (Gossard and Hooke, 1975):

$$\frac{d^2\psi}{dy^2} - \left[k^2 + \frac{d^2U/dy^2}{U-c} - \frac{N^2}{(U-c)^2}\right]\psi = 0. \tag{5.16}$$

Suppose that $U(-\infty) \equiv U_- = 0$, $U(+\infty) \equiv U_0 > 0$ in the shear layer, and the wave is incident from a motionless medium. Then, the solution of (5.16) has the asymptotic form (4.25). The complex coefficients of reflection and refraction, R and T, may be found from the solution of the boundary value problem (5.16), (4.25). The simplest solution to this problem is found for a vortex sheet flow (see Sect. 4.4). An approximate solution can be found for an arbitrarily shaped continuous velocity profile if the problem has a small parameter, i. e., a relative depth of velocity profile $\mu = k_0 l \ll 1$.

It will be shown below in the text that, with the stratification near the critical layer taken into consideration, the corrections to the vortex sheet are proportional to Richardson's number in the critical layer, $\mathrm{Ri}_c \sim N_c^2 l^2/U_0^2$, where N_c is the Brunt–Väisälä frequency in the shear region which generally does not coincide with $N_\pm$. The appropriate corrections in the expressions for R and T are comparable with the corrections emerging in an unstratified medium for a narrow velocity profile (see Sect. 5.1) that have the same order of magnitude for μ only if $\mathrm{Ri}_c \geq \mu$. Evidently, the resonant interaction of IGW with the flow is possible only if the relative phase velocity $\tilde{c} = \omega/kU_0 = N_-/k_0U_0 < 1$ and, consequently,

$$\mathrm{Ri}_c \sim \left(\frac{N}{N_-}\right)^2 \mu^2 \left(\frac{N_-}{k_0U_0}\right)^2 \leq \mu^2 \left(\frac{N}{N_-}\right)^2. \tag{5.17}$$

The condition $\mathrm{Ri}_c \geq \mu$ may be fulfilled in this case if $N^2/N_-^2 \geq \mu^{-1}$, i. e., if the profile $N^2(y)$ has a peak proportional to μ^{-1} in the shear flow region. This means that the density profile (or the profile of potential temperature (Gossard and Hooke, 1975)) has a drop in the shear layer of thickness l. In the absence of such a drop, the effect of stratification may be neglected in a narrow shear flow.

Consider a narrow shear layer for which one has

$$U = U_0\varphi(\xi), \quad N^2 = N_0^2(\xi) + N_1^2(\xi), \quad \xi = y/l,$$
$$f(\xi) = \left(\frac{N_0}{k_0U_0}\right)^2 \sim 1, \quad g(\xi) = k_0 l\left(\frac{N_1}{k_0U_0}\right)^2 \sim 1. \tag{5.18}$$

Let for $y \to \pm\infty$ the parameters of the flow tend exponentially quickly to the corresponding constant values

$$\varphi \to \varphi_\pm \quad (\varphi_+ = 1,\ \varphi_- = 0),$$
$$f \to f_\pm, \quad g \to 0, \quad N_0 \to N_\pm, \quad N_1 \to 0. \tag{5.19}$$

In this case, the expressions (4.25) give an external series expansion to exponential accuracy. By expressing this series expansion through the internal variable ξ one finds

$$\begin{aligned}(\psi_{e1}^+)_{i2} &= \left(1 - i\mu\xi\frac{q_+}{k_0} - \mu^2\xi^2\frac{q_+^2}{2k_0^2}\right)T,\\ (\psi_{e1}^-)_{i2} &= 1 + R - i\mu\xi(1-R)\frac{q_-}{k_0} - \mu^2\xi^2(1+R)\frac{q_-^2}{2k_0^2}.\end{aligned} \tag{5.20}$$

Now let us write (5.16) with the internal variable ξ:

$$\frac{d^2\psi}{d\xi^2} - \frac{\varphi''(\xi)}{\varphi - c}\psi = -\frac{\mu g(\xi)}{(\varphi-c)^2}\psi + \mu^2\left[\sin^2\theta - \frac{f}{(\varphi-c)^2}\right]\psi. \tag{5.21}$$

By employing the method of successive approximations one can readily find the second-order internal expansion

$$\psi_{i2} = A(\varphi - c)\left\{1 - \mu\int_0^\xi \frac{d\xi}{(\varphi-c)^2}\right.$$

$$\times \int_0^{\zeta} \left[g(\chi) \left(1 - \mu \int_0^{\chi} \frac{d\tau}{(\varphi - c)^2} \int_0^{\tau} g(\theta)\, d\theta \right) - \mu[(\varphi - c)^2 \sin^2\theta - f] \right] d\chi \Bigg\}$$

$$+ \mu B \int_0^{\xi} \frac{d\zeta}{(\varphi - c)^2} \left[1 - \mu \int_0^{\zeta} g(\chi) d\chi \int_0^{\chi} \frac{d\tau}{(\varphi - c)^2} \right]. \tag{5.22}$$

Let us express this internal series expansion through the external variable σ retaining the second-order approximation

$$(\psi_{i2})^{\pm}_{e2} = A(\varphi_{\pm} - c) \Bigg\{ 1 - \mu V^{\pm}_{-2} - \mu \frac{\xi D^{\pm} - G^{\pm}}{(\varphi_{\pm} - c)^2}$$

$$+ \frac{\mu^2 \xi}{(\varphi_{\pm} - c)^2} \left[W^{\pm}_{2} + \frac{2M^{\pm} - D^{\pm} G^{\pm}}{(\varphi_{\pm} - c)^2} + D^{\pm} V^{\pm}_{-2} - N^{\pm}_{-2} \right] \Bigg\} + \mu B (\varphi_{\pm} - c) H, \tag{5.23}$$

where $H = W^{\pm}_{-2} + \dfrac{\xi}{(\varphi_{\pm} - c)^2} - \dfrac{\mu\xi \left[G^{\pm} + (\varphi_{\pm} - c)^2 (D^{\pm} W^{\pm}_{-2} - V^{\pm}_{-2}) \right]}{(\varphi_{\pm} - c)^3}$,

and the following notation is used:

$$D^{\pm} = \int_0^{\pm\infty} g(\xi)\, d\xi, \quad G^{\pm} = \int_0^{\pm\infty} \xi g(\xi)\, d\xi, \quad V^{\pm}_{-2} = \int_0^{\pm\infty} \Psi^{\pm}_{-2}\, d\xi \int_0^{\xi} g(\chi)\, d\chi,$$

$$W^{\pm}_{2} = \int_0^{\pm\infty} [\sin^2\theta \Psi^{\pm}_{2} - (f - f_{\pm})]\, d\xi, \quad W^{\pm}_{-2} = \int_0^{\pm\infty} \Psi^{\pm}_{-2}\, d\xi, \tag{5.24}$$

$$M^{\pm} = \int_0^{\pm\infty} \xi g(\xi)\, d\xi \int_0^{\xi} g(\chi)\, d\chi, \quad N^{\pm}_{-2} = \int_0^{\pm\infty} \Psi^{\pm}_{-2}\, d\xi \left(\int_0^{\xi} g(\chi)\, d\chi \right)^2,$$

$$\Psi^{\pm}_{\pm 2} = (\varphi - c)^{\pm 2} - (\varphi_{\pm} - c)^{\pm 2}.$$

Now one can "match" the internal and external series expansions by simply equating (5.23) and (5.20). Note that the coefficients of reflection and refraction may be calculated knowing only the asymptotic forms of the solutions of equation (5.21) at $\xi \to \pm\infty$ where the method of successive approximations holds. The solutions are matched at $\xi \to +\infty$ and $\xi \to -\infty$ across the region containing the singularities of the coefficients in (5.21) (i. e., for $\varphi = c$) employing Lin's rule.

The coefficients R and T found in this fashion enable one to find the value of relative wave amplification

$$\Pi = |R|^2 + |T|^2 \mathrm{Re}\left(\frac{q_+}{q_-}\right) - 1$$

that describes the excess of the total energy flow of the reflected and refracted waves over the unit energy flow of the incident wave:

$$\Pi = -\frac{4\mu q_-}{k_0}\frac{(\varphi_- - c)^2 \mathrm{Im}\,[q_+^2(\varphi_+ - c)^4 W_{-2} + \tilde{N}_{-2}k_0^2]}{[q_-(\varphi_- - c)^2 + q_+(\varphi_+ - c)^2]^2 + Dk_0^2}. \tag{5.25}$$

Here, $W_{-2} = W_{-2}^+ - W_{-2}^-$, $\tilde{N}_{-2} = \int\limits_{-\infty}^{+\infty} \Psi_{-2}^- \, d\xi \left(\int\limits_{\xi}^{+\infty} g(\chi)\, d\chi\right)^2$, and the coordinate origin is chosen so that $D^+ = -D^- = D/2$.

In the absence of refracted wave ($\mathrm{Re}\, q_+ = 0$), the expression for Π takes the form:

$$\Pi = |R|^2 - 1 = -\frac{4\mu q_-}{k_0}(\varphi_- - c)^2$$
$$\times \frac{\mathrm{Im}\left[\tilde{N}_{-2}k_0^2 - 2k_0^2(\varphi_+ - c)^2\tilde{V}_{-2} + |q_+|^2(\varphi_+ - c)^4 W_{-2}\right]}{[q_+(\varphi_+ - c)^2 - Dk_0]^2 + q_-^2(\varphi_- - c)^4}, \tag{5.26}$$

where $\tilde{V}_{-2} = \int\limits_{-\infty}^{+\infty} \Psi_{-2}^- \, d\xi \int\limits_{\xi}^{+\infty} g(\chi)\, d\chi$.

The quantity Π is a measure of resonant wave amplification in the flow. In the limit $\mu \to 0$, when the shear layer transforms to a vortex sheet, Π vanishes, which corresponds to the conservation of the wave energy on reflection at the vortex sheet. When $\mu \neq 0$, Π is determined by the imaginary parts of the integrals W_{-2}, $\tilde{V}_{-2}$ and $\tilde{N}_{-2}$ that must be calculated making allowance for Lin's rule for the poles $\varphi(\xi_c) = c$:

$$\operatorname{Im}(W_{-2}) = -\pi \left\{ \frac{\varphi''(\xi)}{[\varphi'(\xi)]^2} \right\}_{\xi=\xi_c},$$

$$\operatorname{Im}(\tilde{V}_{-2}) = -\pi \left\{ \frac{\varphi''(\xi)}{[\varphi'(\xi)]^2} \int_{\xi}^{\infty} g(\chi)\, d\chi + \frac{g(\xi)}{[\varphi'(\xi)]^2} \right\}_{\xi=\xi_c}, \tag{5.27}$$

$$\operatorname{Im}(\tilde{N}_{-2}) = -\pi \left\{ \frac{\varphi''(\xi)}{[\varphi'(\xi)]^2} \left(\int_{\xi}^{\infty} g(\chi)\, d\chi \right)^2 + \frac{2g(\xi)}{[\varphi'(\xi)]^2} \int_{\xi}^{\infty} g(\chi)\, d\chi \right\}_{\xi=\xi_c},$$

where the primes denote the derivatives with respect to the corresponding variables.

With (5.27) taken into account, the expression for the relative amplification, finally, takes the form

$$\Pi = \frac{4\pi |q_- q_+^2| c^2 (U-c)^4 L}{[q_- c^2 + q_+(U_0 - c)^2]^2 + I^2}, \tag{5.28}$$

where $L = \left[\frac{1}{U'(y)} \frac{d}{dy} \frac{|p(y)/p(\infty)|^2}{U'(y)} \right]_{y=y_c}$, $\quad I = \int_{-\infty}^{+\infty} N_1^2\, dy.$

This expression includes the asymptotic series expansion of the pressure amplitude in the first-order approximation with respect to μ:

$$\frac{p(y)}{p(\infty)} = 1 - \frac{i}{q_+(U_0 - c)^2} \int_{y}^{\infty} N_1^2\, dy = 1 + \frac{\rho_0 \eta}{p(\infty)} \int_{y}^{\infty} N_1^2\, dy$$

(η being the amplitude of the displacement of the isopycnic surface) that can be obtained by substituting (5.22) into the expression for pressure:

$$p = -\rho_0 (U-c)^2 \frac{d}{dy} \frac{\psi}{U-c}. \tag{5.29}$$

So, a resonant interaction of a wave with a flow in a critical layer leads to wave amplification at $\Pi > 0$.

5.2.1 The criterion of resonant amplification. Physical interpretation

The condition of internal wave amplification caused by its reflection at a shear flow can be represented in the form:

$$\left[\frac{1}{U'(y)}\frac{d}{dy}\left(\frac{1}{U'(y)}\left|\frac{p(y)}{p(\infty)}\right|^2\right)\right]_{y=y_c} > 0. \tag{5.30}$$

This criterion has a simple physical meaning if one takes into account that, according to what was said in Sect. 2.1, the expression $f(y) = 1/U'(y) > 0$ is analogous to the function of particle velocity distribution. In an unstratified critical layer one has $d|p|^2/dy = 0$. In this case, the amplification is proportional to the derivative $df/dU = -U''(y)/U'(y)^3$ calculated at the critical point y_c. It is common knowledge that the direction of the energy exchange between the wave and flow in a plasma is determined by the difference of the amount of the particles going ahead of the wave and lagging behind it in the neighborhood of resonance. The plasma–hydrodynamic analogy — the similarity of the resonant interaction in hydrodynamical flows and the kinetic resonance effects such as Landau damping in a plasma — were discussed in Sect. 2.1.

In the presence of pronounced stratification ($N_1 \neq 0$, $\mathrm{Ri}_c \geq \mu$), the wave field intensity (in the first-order approximation with respect to μ) becomes inhomogeneous: $(d|p|^2/dy)_{y=y_c} \neq 0$. Evidently, this inhomogeneity depends on the gradient of the amplitude of oscillations of the hydrostatic pressure arising as the liquid layer is periodically shifted with variable density:

$$p = \Delta\rho_0 g\eta = \rho_0\eta\int_y^\infty N_1^2\,dy,$$

where η is a perturbation of the isopycnic surface. Because of the inhomogeneity of the wave field, the interaction with the particles going ahead and lagging behind occurs with different intensities. Now the direction of energy exchange is determined, according to (5.30), by the arbitrary quantity $|p|^2 f(U)$.

It should be noted that the interaction of perturbations with a stratified flow is frequently treated incorrectly. Within this interpretation the amplification of the wave perturbations is attributed to the decrease of the kinetic energy of the flow in the neighborhood of a critical layer and their absorption, to the increase of the potential energy of the particles of the flow (the heavier particles lift to the surface and the lighter ones lower in a stable stratified medium). The relationship between the amplification and absorption are then described by the Richardson number at a critical point. Actually, the example of the flow with a linear velocity profile where $df/dU = 0$ demonstrates that IGWs may be either amplified or absorbed at the same properties of the flow in a critical layer. The true criterion of amplification has the form

$$|q_+|(U_0 - c)^2 > \int_{y_c}^{\infty} N_1^2 \, dy. \tag{5.31}$$

At the same time, by averaging the continuity equation in the Boussinesq approximation:

$$\frac{\partial < \rho >}{\partial t} + \frac{\partial < \tilde{\rho}\tilde{v} >}{\partial y} = 0, \quad \tilde{\rho} = -\frac{i\rho_0 N^2 \tilde{v}}{g(\omega - kU)}$$

($\tilde{\rho}$ is the amplitude of density perturbation) one can show (Tsimring, 1982) that $\partial\langle\rho\rangle/\partial t = 0$. Then, the potential energy of the particles of the liquid in a stratified flow does not change under the action of IGWs.

The interpretation of resonance interaction in terms of the competition between the absorption and amplification when the perturbation is affected by the particles going ahead and lagging behind the flow explains the well-known effect of IGW absorption in the critical layer with a large Richardson number. The wave is absorbed by the flow "particles" (jets) while only a small portion of the wave field reaches the particles that go ahead.

Since the wave amplification and absorption depend on the wave field in the critical layer, the criterion of amplification cannot be reduced to the local condition such as the Miles criterion $U''(y_c) < 0$. The IGW amplification in a stratified flow is determined by the global condition, i. e., it depends not only on the details of the flow profiles at the critical point but also on the properties of the wave field the behavior of which at the critical point depends on the entire flow and on boundary conditions as $y \to \infty$.

One can obtain a more general, compared to (5.30), amplification condition that is valid for arbitrary Richardson numbers. For this purpose it suffices to pass from (5.16) to the equation for pressure by substituting the variable (5.29):

$$\frac{d}{dy}\frac{dp/dy}{(\omega - kU)^2 - N^2} - \frac{k^2 p}{(\omega - kU)^2} = 0. \tag{5.32}$$

Let us multiply (5.32) by p^*, integrate the found equality with respect to y, and separate the imaginary part specified by the pole y_c:

$$\mathrm{Im}\left[\frac{p^* dp/dy}{(U-c)^2 - N^2/k^2}\right]_{-\infty}^{+\infty} = \mathrm{Im}\int_{-\infty}^{+\infty} \frac{k^2|p|^2}{(U-c)^2}\, dy. \tag{5.33}$$

Now the substitution of the boundary conditions for pressure amplitude at $y = \pm\infty$ yields the amplification condition

$$\Pi = -\frac{1}{\rho_0 |q_-|}\mathrm{Im}\int_{-\infty}^{+\infty} \frac{k^2|p|^2}{(U-c)^2}\, dy > 0. \tag{5.34}$$

The criterion (5.30) which was obtained above by the method of matched asymptotic series expansions for $\mu \ll 1$ follows from (5.34), given that the integrand contains the asymptotic expression to the first-order approximation with respect to μ for the pressure amplitude without singularities at the point y_c, and calculations are performed in the upper half-plane of the complex plane $\omega = \omega_r + i\delta$, which agrees with the solution of the initial problem by the Laplace method. Note that the exact expression for pressure amplitude has a singularity at $y = y_c$. This does not allow one to shift the integration path in (5.34) to the complex plane y, i. e., to reduce the integration in (5.34) at $\delta \to 0$ merely to Lin's integration rule.

The condition (5.34) for arbitrary flows may be written in a different form. The imaginary part of the integral is, evidently, determined by the behavior of the function $p(y)$ in the vicinity of the point $y = y_c$. The solution of (5.16) near the critical point at Ri < 1/4 has the form (Gossard and Hooke,1975):

$$\psi = A(y - y_c)^{1/2+\nu} + B(y - y_c)^{1/2-\nu}, \tag{5.35}$$

where $\nu = \sqrt{1/4 - \mathrm{Ri}_c}$. Let us now substitute (5.35) into the expression (5.29) for pressure and shift the critical point, in accord with Lin's rule, to the upper half-plane of complex y (for $dU/dy > 0$), i. e., let us set $y_c = y_0 + i\Delta$ ($\Delta > 0$). Then, direct calculation of the integral (5.34) gives the well-known criterion of IGW amplification by a flow (Jones, 1968):

$$\mathrm{Im}\,(A^* B e^{i\pi\nu}) < 0. \tag{5.36}$$

This criterion, similarly to (5.34), is valid for arbitrary Richardson numbers $\mathrm{Ri}_c < 1/4$. Obviously, the general criterion of amplification cannot be represented explicitly, i. e., directly through the parameters of the flow. However, it is useful, as an illustration, to find an analytical criterion for some particular flows.

5.2.2 Analytical and numerical investigations. Spontaneous radiation by a critical layer

The simplest example of a stratified flow with a critical layer is a flow with a linear velocity profile

$$U(y) = \begin{cases} -U_0 \equiv -\alpha l; \\ \alpha y; \\ U_0 \equiv \alpha l; \end{cases} \qquad N^2 = \begin{cases} N_0^2 & \text{at } y < -l; \\ N_0^2 + N_1^2 & \text{at } |y| < l; \\ N_0^2 & \text{at } y > l. \end{cases} \tag{5.37}$$

In this case, the solution to (5.16) is expressed through the modified Bessel function

$$\psi = (y - y_c)^{1/2}\{AI_\nu[k(y-y_c)] + BI_{-\nu}[k(y-y_c)]\}. \qquad (5.38)$$

Using the boundary conditions at the velocity kink ($y = \pm l$), the pressure and stream function continuity, one can readily obtain from (5.36) the amplification condition in the form

$$\mathrm{Im}\left[\frac{\delta F_{-\nu} + i\delta q_+/k - 1/2}{-\delta F_\nu + i\delta q_+/k + 1/2}\frac{I_{-\nu}}{I_\nu}e^{i\pi\nu}\right] > 0, \qquad (5.39)$$

where $\delta = k(l - y_c)$ and $F_{\pm\nu} = I'_{\pm\nu}(\delta)/I_{\pm\nu}(\delta)$. The simplest result is found for $\delta \ll 1$ and $\mathrm{Ri}_+ = (N_+/\alpha)^2 < \delta^2$. There is norefracted wave in this case and $iq_+ = -|q_+|$, while the condition (5.39) takes the form

$$|q_+|(l - y_c) > \mathrm{Ri}_c. \qquad (5.40)$$

One can easily see that for the flow (5.37) the condition (5.31) simplifies to ($U''(y) = 0$) and reduces to (5.40).

Consider still another example of the velocity profile that has an analytical solution at $N = \text{const}$ in the form of a parabolic profile

$$U(y) = U_0(1 - y^2/L^2), \qquad (5.41)$$

given that the condition of the "rigid top" $\psi(-d) = 0$ is satisfied at the point $y = -d$ (instead of the condition (5.28) at $y \to \infty$), and the wavelength is large so that $kL \ll 1$. For $k = 0$, the solution to (5.16) with the profile (5.41) has the following form

$$\begin{aligned}\psi = {}& A(y + |y_c|)^{1/2+\nu}(y - |y_c|)^{1/2-\nu}(y - 2\nu|y_c|) \\ &+ B(y + |y_c|)^{1/2-\nu}(y - |y_c|)^{1/2+\nu}(y + 2\nu|y_c|), \end{aligned} \qquad (5.42)$$

where $y_c = -L(1 - c/U_0)^{1/2}$. By substituting this solution into the boundary condition at $y = -d$ one obtains the A/B ratio. As a consequence, the criterion (5.36) takes on a form

$$\frac{1}{y_c} = -\left[\frac{U''(y)}{U'(y)}\right]_{y=y_c} > \frac{1 - \sqrt{1 - 4\mathrm{Ri}_c}}{l}. \qquad (5.43)$$

Here, $l \equiv |y_c| - d$ is the distance from the critical layer to the rigid top.

Note that the criteria (5.40) and (5.43) are nonlocal in the two considered examples: the condition of IGW amplification depends not only on the features of the flow near the resonance but also on the geometry of the shear layer as a whole.

Consider now results of the numerical calculation of the problem (5.16), (4.25) for the shear layer with the velocity profile (5.15). The magnitude of the reflection coefficients is determined in this case by two parameters of the wave: its phase velocity c and the angle of incidence θ as well as by the parameter of the flow, i. e., by the Richardson number at the inflection point $\mathrm{Ri} = 4(Nl/U_0)^2$.

The angular dependences of the coefficients R and T were calculated at different parameters (see Fig. 5.2, where the values of c are indicated on the corresponding curves) and were used to find the angular dependences of the relative amplification $\Pi(\theta)$ (Fig. 5.3).

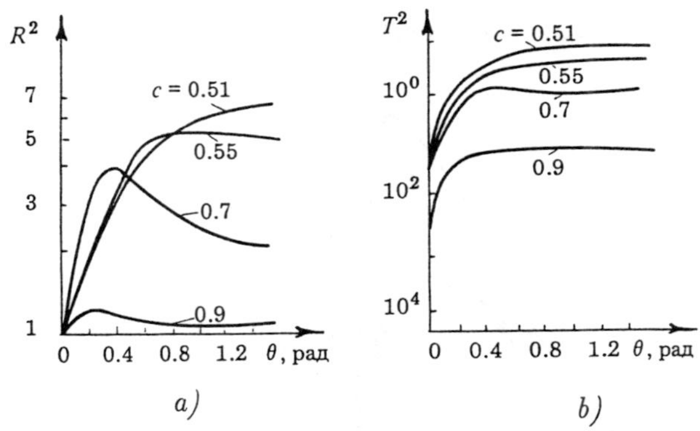

Figure 5.2: Angular dependences of the reflection and refraction coefficients of IGW at Ri = 0.012 and different values of relative phase velocity $v = \omega - kU_0$.

Note that the resonance amplification here is possible not only for $c > 0.5$, when the critical point is in the convex portion of the profile, but also for $c < 0.5$, when the waves are in resonance with the concave portions of the profile. A typical feature of $R(\theta)$ and $\Pi(\theta)$ at $c < 0.5$ is the presence of a sharp maximum in the narrow angular range.

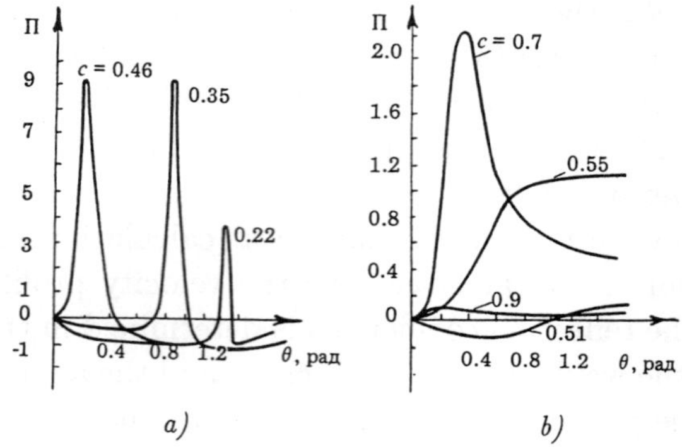

Figure 5.3: Angular dependences of relative IGW amplification: (a) Ri $= 0.1$, $c < 0.5$; (b) Ri $= 0.012$, $c > 0.5$.

The maximal values of $\Pi_{\max}(\mathrm{Ri}, c)$ were determined for each plot of $\Pi(\theta)$ drawn for fixed Ri and c. The isolines $\Pi_{\max} = \mathrm{const}$ in the plane of the parameters (Ri, c) are shown in Fig. 5.4.

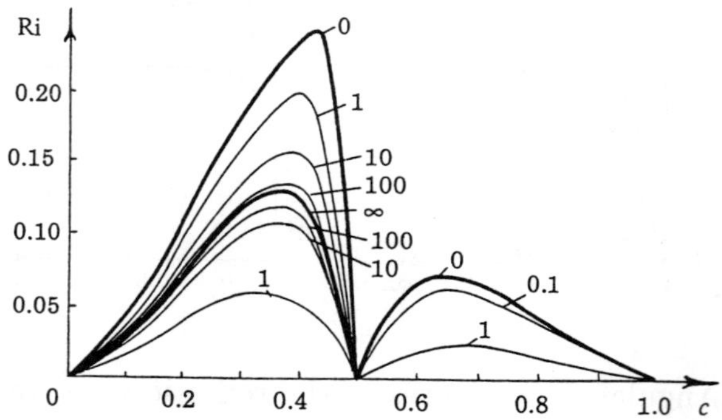

Figure 5.4: Isolines $\Pi_{\max} = \mathrm{const}$.

An important feature here is the presence of the isoline $\Pi_{max} = \infty$ that exists at $c < 0.5$. It corresponds to the resonance between the incident wave and the neutral eigenmode of the system in which the field is oscillating below the shear layer and is decaying exponentially above it. Such neutral perturbations which correspond to the spontaneous radiation of IGW by the shear layer were found by Kuramoto and Tsuzuki (1976) and by Fritts (1980). The isolines $\Pi_{\max} = \text{const}$ in Fig. 5.5 are depicted in the plane of the parameters (Ri, kl). The isoline $\Pi_{\max} = \infty$ coincides here with one of the neutral curves in the same plane of the parameters as were obtained by Fritts (1980).

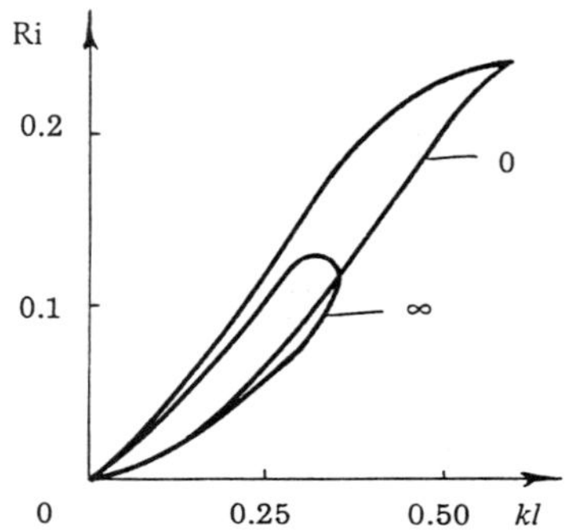

Figure 5.5: Isolines $\Pi_{\max} = 0, \infty$ in the plane (kl, Ri).

Note that, unlike the case of spontaneous radiation by a vortex sheet, the IGW energy is not conserved in the case considered here: there exists a critical layer in which the wave draws the energy from the stratified flow.

5.3 Internal waves in horizontally inhomogeneous shear flows

The spatial structure of the flows in the atmosphere and in the ocean typically has both vertical and horizontal inhomogeneities. The horizontal gradients of the mean density and flow fields are usually much smaller than the vertical ones. However, the group velocity of IGWs is most often directed at a small angle to the horizon and the IGW packets may propagate hundreds and thousands of kilometers (Francis, 1975). It is necessary to take into account the effect of horizontal inhomogeneity of the medium at such distances.

The equation for the small pressure perturbations in a shear flow of an incompressible stratified liquid with the velocity $U(y,z)$ directed along the x-axis and density $\rho_0(y,z)$ is written in the Boussinesq approximation as

$$\frac{\Omega^2}{k_x^2}\frac{\partial}{\partial y}\left(\frac{\Omega^2}{k_x^2}\frac{\partial p}{\partial y}\right)+\frac{\partial}{\partial z}\left[B\left(\frac{\partial p}{\partial z}+\frac{g}{\rho_0\Omega^2}\frac{\partial p}{\partial y}\frac{\partial \rho_0}{\partial y}\right)\right]+k_x^2 p=0, \quad (5.44)$$

where $B=\left[1-N^2(y,z)/\Omega^2\right]^{-1}$ and $\Omega=\omega-k_xU$ is a wave frequency in a moving reference system.

In the vertically inhomogeneous flow with the profile $U(z)$ and exponential density distribution ($N=\text{const}$), one can obtain from (5.44) the Taylor–Goldstein equation (5.16) in which the vertical coordinate z is an independent variable. IGWs in a horizontally inhomogeneous shear flow were investigated by Basovich and Tsimring (1984) in the frame of a simple model with the Brunt–Väisälä frequency and flow velocity constant in depth. For the pressure perturbations proportional to $\exp(-i\omega t+ik_xx+iqz)$ in the flow with the velocity profile $U(y)$, (5.44) gives

$$\frac{d^2p}{dy^2}+\frac{2U''(y)k_x}{\Omega}\frac{dp}{dy}+\left(\frac{\Omega^2}{N^2-\Omega^2}q^2-k_x^2\right)p=0. \quad (5.45)$$

By employing the WKB method for the flows in which the inhomogeneity scale is larger than the length of the internal wave one can obtain approximate solutions to (5.45). The substitution of the variable $p=-\Omega b/k_x$ transforms this equation to the canonical form

$$\frac{d^2b}{dy^2}+K^2(y)b=0, \quad (5.46)$$

where $K^2(y)=n^2+\dfrac{\Omega''(y)}{\Omega(y)}-2\left[\dfrac{\Omega'(y)}{\Omega(y)}\right]^2,\quad n^2=\dfrac{\Omega^2q^2}{N^2-\Omega^2}-k_x^2.$

Equation (5.46) is analogous to the Schrödinger equation for the wave function of the particle in the potential $-K^2(y)$ that is well-known in quantum mechanics. This equationhas two linearly independent WKB solutions:

$$b_{1,2} = \left[K^2(y)\right]^{-1/4} \exp\left\{\pm i \int |K^2(y)|\, dy\right\}. \qquad (5.47)$$

The potential relief $-K^2(y)$ for the shear flow with the profile (5.15) is shown in Fig. 5.6 where it is assumed that N = const, $k_x > 0$, $U_0 > (\omega + N)/k_x$, and $\omega < N$.

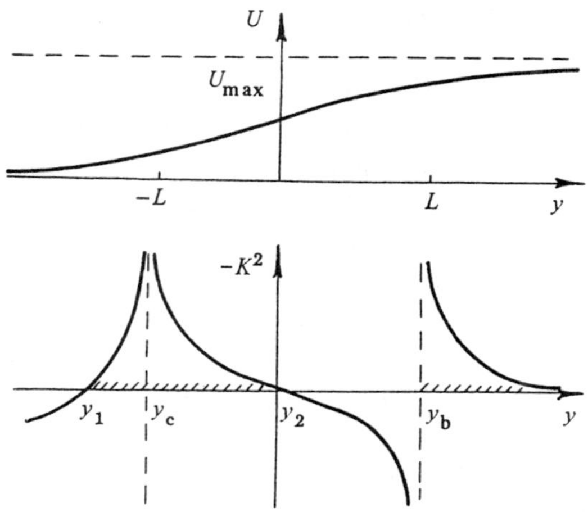

Figure 5.6: Potential relief $-K^2(y)$ for the IGW in a shear flow. Hatched are the opaque regions.

The potential relief contains the singular points: the resonance singularity y_c in the critical layer where $U(y_c) = \omega/k_x$ and the limiting points y_b at which $U(y_b) = (\omega + N)/k_x$. The WKB approximation does not hold in the neighborhood of these points and near the turning point where $K(y) = 0$.

As is well seen from Fig. 5.6, the waves running against the shear flow are reflected at the turning point y_1. The nontransmitting region $y_1 < y < y_2$ contains the resonance singular point y_c. The wave partially penetrates through the potential barrier formed by the nontransmitting region and gets to the region of propagation $y_2 < y < y_b$ that is bounded by the limiting layer y_b. The internal waves with the corresponding ω and k_x cannot propagate at the flow velocities $U(y) > U(y_b)$, i. e., at $y > y_b$ because the absolute value of the wave frequency in the co-moving reference system $\Omega = \omega - k_x U$ exceeds the Brunt–Väisälä frequency $|\Omega| > N$.

The behavior of the solutions at singular points specifies the coefficient of wave reflection at the potential barrier. For the calculation of the reflection coefficient one needs to construct the asymptotic form of the solution to (5.45) throughout the region y by matching the WKB solutions (5.46) at the turning points y_1 and y_2 and at the singular points y_c and y_b taking into account the boundary condition at $y \to \infty$. The solutions are matched at the turning points in a standard fashion. In order to determine the rules of matching at the points y_c and y_b one can use the reference equations which represent the principal singularities of the behavior of solutions to (5.46) in the neighborhood of these points.

Seeking the solution in the neighborhood of the point y_b let us substitute in the potential $K(y)$ only the first term of the Taylor series expansion of the velocity-profile function

$$U(y) = U'(y_b)(y - y_b) + O[(y - y_b)^2]. \tag{5.48}$$

Equation (5.46) in the neighborhood of the point $y = y_b$ may be written approximately in the form

$$\frac{d^2 b}{dy^2} - \frac{A}{y - y_b} b = 0, \quad \text{where } A = \frac{N k_y^2}{2 k_x U'(y_b)}. \tag{5.49}$$

Such an equation emerges in a great number of problems related to the wave propagation in plane-parallel media (Ginzburg, 1967; Timofeev, 1971). Its solution may be expressed by the first-order Bessel function Z_1:

$$b = b_0 \sqrt{y - y_b} Z_1 \left[2i\sqrt{A(y - y_b)} \right]. \tag{5.50}$$

It is well-known that this function has a branching point at $y = y_b$ (Janke *et al.*, 1960). The choice of the corresponding branch of the solution is governed here by Lin's rule. By using this rule one can find that the exponent damping at $y > y_b$ corresponds to the progressive wave with negative phase velocity at $y < y_b$:

$$\sqrt[4]{y - y_b}\exp\left[-2\sqrt{A(y - y_b)}\right] \longleftrightarrow \sqrt[4]{y_b - y}\exp\left[-2i\sqrt{A(y - y_b)}\right].$$

This wave possesses negative energy because $\Omega = \omega - k_x U < 0$ in the region $y > y_c$ and has opposite projections of the phase and group velocities onto the y-axis:

$$c_y (c_g)_y \equiv \frac{\omega}{k_y}\left(\frac{d\omega}{dk}\right)_y = \frac{\omega}{\omega - k_x U}\frac{q^2 N^2}{(k_x^2 + k_y^2 + q^2)^2} < 0. \tag{5.51}$$

The group velocity at $y > y_2$ is directed towards the limiting point. There is no reflected wave in this case, hence, short IGWs are totally absorbed at the limiting point. There naturally arises a question of the origin of the effect of total wave absorption.

A paradoxical situation occurs, at first sight, in a perfect fluid where there is no dissipation. The explanation of this effect is rather transparent: the approximation of a stationary wave process does not hold in the neighborhood of the limiting point. Indeed, an infinite energy must be accumulated near $y = y_b$ whereas the wave energy flux is finite. Therefore, one must consider an intermediate process of energy accumulation in the neighborhood of the limiting point. A bounded internal wave train approaches asymptotically the limiting point, with the wave energy density increasing until dissipation or nonlinear effects appear.

If the velocity U_0 were sufficiently small so that the limiting point y_b were absent (but the point y_2 and the transparency region past it existed), the solution would be constructed proceeding from the condition of radiation at $y > y_2$. Then, the wave transmitted through the barrier (y_1, y_2) would have the group velocity $(c_g)_y > 0$. A similar procedure can be employed in the presence of a limiting layer because it completely absorbs the wave incident on it and there is no reflected wave. The WKB solutions in the vicinity of the turning points y_1 and y_2 are matched by means of the Airy function. The solutions in the neighborhood of the resonance singular point y_c are related by the reference equation the solutions to which are the Wittaker functions. The corresponding branches of these functions are chosen on the basis of Lin's rule.

By solving the boundary value problem in this fashion one can find the reflection coefficient for short IGWs in the first-order approximation with respect to the small parameter $\mu^{-1} = (kL)^{-1}$. The value of the amplification of the reflected wave is described by the following expression

$$|R|^2 - 1 = \exp(-2\delta) + \frac{\pi}{k}\frac{U''(y_c)}{|U'(y_c)|}\exp(-2\delta_1), \tag{5.52}$$

where $\delta = i\int\limits_{y_1}^{y_2} n(y)\,dy, \quad \delta_1 = i\int\limits_{y_1}^{y_c} n(y)\,dy.$

The absolute value of the reflection coefficient may be both greater and lesser than unity, depending on the ratio of two terms in the right-hand side of (5.51). The first term is always positive and is closely related to the excitation past the potential barrier of the negative energy wave which is then absorbed at the limiting point or vanishes to the infinity. The interaction with the refracted NEW gives rise to amplification of the reflected wave.

The second term in (5.51) is determined by the resonant interaction of the wave with the flow in the critical layer. Its sign depends on the curvature of the velocity profile, similarly to the analogous problem for a sound wave (see Sect. 5.1). Note that the magnitude of wave amplification (or absorption) in the WKB approximation is exponentially small because the critical level y_c and the NEW region are past the potential barrier behind which the field decreases exponentially.

Localized IGW eigenmodes (Fig. 5.7a) may exist in a jet flow if the maximal velocity of the wave does not exceed $(\omega + N)/k_x$. The localized waves in the jet may become unstable due to the amplification mechanisms considered above, i. e., radiative instability and resonant interaction with the flow.

The potential relief changes qualitatively (Fig. 5.7b) for a sufficiently

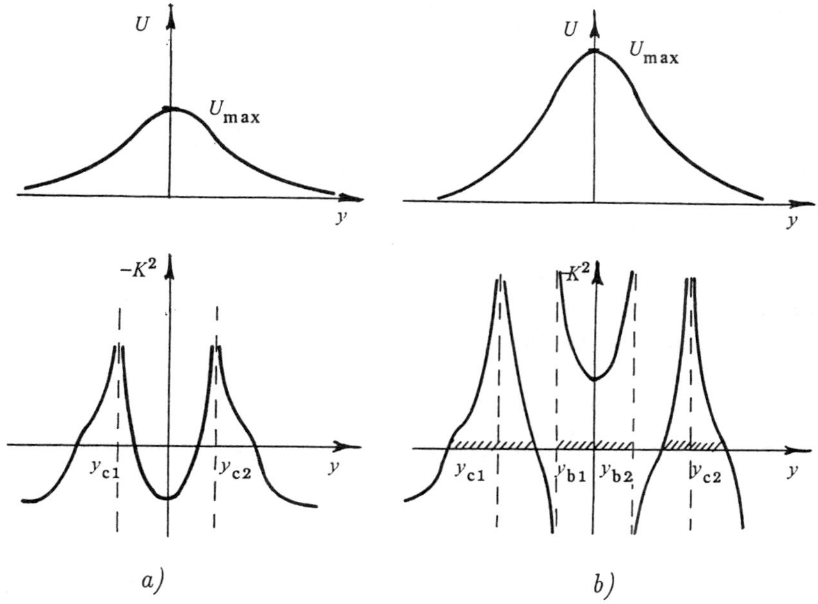

Figure 5.7: Potential relief $-K^2(y)$ for the IGW in a jet flow:
(a) $U_m < (\omega + N)/k_x$, no limiting layers;
(b) $U_m > (\omega + N)/k_x$, there exist the limiting layers $y = y_b$.

fast jet or if the waves are short enough, when $k_x U_0 > \omega + U_0$. Two limiting layers are formed in the neighborhood of the jet center and the region between them is nontransparent for the waves with the given parameters. In this situation no neutral or unstable eigenmodes exist.

5.3.1 Vertical focusing of internal waves

It was mentioned above that internal gravity waves may be trapped by large-scale horizontal inhomogeneities of background fields. The phenomenon of trapping consists in the following. The wave with the intrinsic frequency Ω (in the reference system co-moving with the flow) asymptotically approaches the layer where the frequency Ω equals the local value of the Brunt–Väisälä frequency N which is supposed to be constant throughout the depth. As the monochromatic wave is approaching the trapping layer, its amplitude tends to the infinity.

However, the trapping is not a rough effect in the case of exponential stratification. Even a slight vertical displacement of the background flow velocity leads to wave reflection at the layer $\Omega = N$ (Olbers, 1981). Thus, the simple model considered above is insufficient for adequate description of IGW propagation in an inhomogeneous flow. It is also essential to take into account both the horizontal and the vertical gradients of velocity and of the parameters of the medium.

In a general formulation, the description of the IGW dynamics in an inhomogeneous flow is a rather complicated problem even in a linear approximation. One cannot, in general, separate the variables in the frame of the partial differential equation (5.43) taking into account the vertical and horizontal inhomogeneities simultaneously. Therefore, the IGW evolution in a medium where the horizontal inhomogeneity is much smoother than the vertical one is usually described by a modified WKB method for multimode systems (Badulin *et al.*, 1985). This method allows for approximate separation of variables and reduces the problem to separate analysis of the vertical mode structure and the slow evolution of the wave parameters along the horizontal coordinates.

Taking the ratio of the wavelength to the characteristic scale of horizontal inhomogeneity as a small parameter, (5.43) gives, to a zero or adiabatic approximation, the following equation:

$$\frac{\partial^2 w}{\partial z^2} - \left\{1 - \frac{N^2(y,z)}{[\omega - k_{\mathrm{x}} U(y,z)]^2}\right\} \left[k_{\mathrm{x}}^2 + k_{\mathrm{y}}^2(y)\right] w = 0 \qquad (5.53)$$

that is the generalized Taylor–Goldstein equation (5.16) for a weakly inhomogeneous medium. The functions U and N in this equation depend on the y-coordinate as on parameter.

It is clear from (5.53) that the levels where $\omega - k_x U = N$ correspond to the turning points separating the region of propagation from the nontransmitting regions. If the profile of the effective Brunt–Väisälä frequency

$$N_{eff} = \frac{N(y,z)}{1 - k_x U(y,z)/\omega}$$

has the maximum at the level z_m:

$$N_{eff}(z_m) = \frac{N_m(y)}{1 - k_x U_m(y)/\omega},$$

then it makes a waveguide for the IGW. The eigenmodes of this waveguide are localized in the near vicinity of the maximum N_{eff} in the neighborhood of the trapping point. Then, one can use the series expansion

$$N_{eff}^2 \approx N_m^2 \left[1 - (z - z_m)^2/d^2\right]. \tag{5.54}$$

Such a series expansion holds for any smooth profile $N_{eff}(z)$. Equation (5.53) transforms in this case to

$$\frac{\partial^2 w}{\partial z^2} - \left[1 - \frac{N_m^2}{\omega^2} + \frac{N_m^2}{\omega^2}\frac{(z - z_m)^2}{d^2}\right](k_x^2 + k_y^2)w = 0. \tag{5.55}$$

The boundary value problem that defines the eigenmode may be easily solved if one takes into account that $\omega \to 0$ at $|z - z_m| \to \infty$ may be taken as boundary conditions due to strong mode localization in the neighborhood of $z = z_m$. The solution of this boundary value problem is expressed by the Hermitian polynomial H_n:

$$w_n(y, z) = H_n[\kappa_n(y)z] \exp\left[-\kappa_n(y)z/2\right], \tag{5.56}$$

where $\kappa_n(y) = \sqrt{k_n N_m/\omega}$, and $k_n(y)$ is the eigenvalue of the corresponding boundary value problem:

$$k_n(y) = \frac{2n+1}{d}\frac{\omega N_m}{N_m^2 - \omega^2}. \tag{5.57}$$

For the fundamental mode ($n = 0$) one has

$$k_0(y) = \frac{1}{d}\frac{\omega N_m}{N_m^2 - \omega^2}, \quad w_0(z, y) = \frac{1}{\sqrt{\pi}} \exp\left[-\frac{k_0 N_m}{2\omega d} z^2\right]. \tag{5.58}$$

In the case of the slow variation of waveguide parameters along the horizontal coordinate y and the adiabatic restructuring of the waveguide eigenmode, the dispersion equation for the eigenmode has the form similar to that for a medium at rest with a constant Brunt–Väisälä frequency:

$$\omega^2 = N_m^2(y)\frac{k_x^2 + k_y^2(y)}{k_x^2 + k_y^2 + \tilde{q}^2}, \tag{5.59}$$

where $\tilde{q}$ is the wavenumber that is the measure of the vertical scale of the eigenmode. However, all the quantities in (5.59) depend on y as on parameter. The use of the adiabatic WKB approximation, i. e., of the dispersion relation for IGW and the mode structure which depend parametrically on y, means extension to the smoothly inhomogeneous case of the relations of the IGW parameters that are well-known for horizontally homogeneous fluids.

The trapping of short IGWs in a horizontally inhomogeneous waveguide is accompanied by focusing along the vertical. Consider as an illustration of this effect the picture of wave rays. The ray paths for IGW and the isolines of N_{eff} are depicted in Fig. 5.8. On IGW reflection at the turning points where $\Omega = N$, the wave propagates almost vertically. The deviation of the rays from the vertical direction is determined by the angle $\theta = \arcsin(\omega/N_{eff})$. The maximal deviation along the ray path between two reflections decreases with decreasing distance to the limiting point at $N_{eff} \to \omega$. As a result, the width of the "zigzag", i. e., of the horizontal path of the ray between two reflections, decreases in the neighborhood of the limiting point.

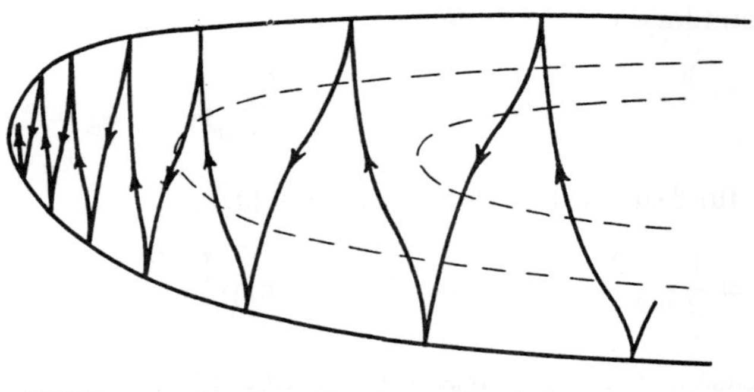

Figure 5.8: IGW trapping in a horizontally inhomogeneous waveguide. The dashed curves stand for the isolines $N_{eff} = \text{const}$.

Let us now demonstrate using the mode language that, for the monochromatic wave with frequency ω, the limiting point at which ω coincides with the local maximum of the effective Brunt–Väisälä frequency N_m is the point of trapping. Let a zero value of y correspond to this point. By expanding ω^2/N_m^2 in a power series of y one obtains the dispersion equation

$$\omega^2 = N_m^2(1 - \alpha y). \tag{5.60}$$

It follows from (5.56), (5.58) that with decreasing distance to the limiting layer, the horizontal wavenumber k_y increases as y^{-1} and the vertical scale of the mode decreases as $y^{1/2}$. Using the expressions (5.58) and (5.59), one can calculate the time τ^* in which the narrow IGW train may reach the point y_* in the neighborhood of the trapping point $y = 0$:

$$\tau^* = \int\limits_{y_0}^{y_*} \frac{dy}{[c_g(y)]_y} = \frac{k_* - k_0}{N\alpha}, \tag{5.61}$$

where k_* is the value of the horizontal wavenumber at the point y_* and k_0 is the initial value of the wavenumber of the IGW train. It follows from (5.60) that the time needed for the IGW train to reach the point $y = 0$ is infinite and, consequently, $N_m = \omega$ is the trapping layer. This significant circumstance is closely related to the possible manifestation of strong dissipative, dispersive, and nonlinear effects in the neighborhood of such a layer.

It is worth noting that the asymptotic behavior of the wave field in the neighborhood of the trapping point may be found avoiding the use of the WKB approximation. When (5.53) and (5.59) are approximated, the partial differential equation (5.43) has exact solutions that are expressed by the hypergeometric functions of complex argument. The relevant problem was solved by Erokhin and Sagdeev (1985) who also obtained the solutions to a two-dimensional problem, with the effects of viscosity and nonstationarity taken into account.

Note also that the effects of linear focusing and trapping were investigated for Langmuir waves in an inhomogeneous magnetized plasma (Erokhin and Moiseev, 1979; Andronov and Chugunov, 1975) the dispersive properties of which are very similar to the properties of IGWs (Golubyatnikov *et al.*, 1960).

5.4 Rossby waves in a critical layer

The analysis described above may also be performed for the Rossby waves. Such an investigation reveals the effects that are very similar to the corresponding effects for IGWs, namely, resonant interaction in the critical layer of a flow with vertical or horizontal displacement, focusing and trapping at a limiting point, etc. However, there also exist pronounced difficulties.

The amplitude of the harmonic perturbations of stream function in a barotropic zonal flow with velocity profile $U(y)$ on a β-plane obeys the following equation

$$\frac{d^2\psi}{dy^2} - k_x^2\psi + \frac{\beta - U''(y)}{U - c}\psi = 0. \tag{5.62}$$

Here, x is the zonal (west–east) coordinate and y is the latitude (south–north) coordinate. For $U = 0$, (5.62) describes the Rossby wave with the wave vector $\mathbf{k} = (k_x, q)$ and gives the dispersion relation (4.33).

The equation for the stream function amplitude in a stratified flow with vertical displacement (baroclinic flow):

$$\frac{d^2\psi}{dz^2} - Mk_x^2\psi + \frac{M\beta - U''(z)}{U - c}\psi = 0 \tag{5.63}$$

differs from (5.62) only by the parameter $M \equiv N^2/(2\Omega_0)^2$ that describes the ratio of the buoyancy force to the Coriolis force (here, Ω_0 is the angular velocity of rotation of the β-plane). Thus, in the frame of (5.62) one can investigate the interaction of the Rossby waves with both barotropic and baroclinic flows, because equation (5.63) is reduced to (5.62) by means of simple transformation $y = z\sqrt{M}$.

Equation (5.62) together with the boundary conditions (4.34) constitute a boundary value problem the solution to which gives the complex coefficients of reflection R and refraction T of a plane wave. By multiplying (5.62) by ψ^* and integrating with respect to the y-coordinate one can show that, by analogy with other types of waves, the quantity $S_y = \mathrm{Re}\,\langle uv^* \rangle = -k\mathrm{Im}\,(\psi^* \psi'_y)$ remains constant along the y-coordinate everywhere except the resonance points y_c at which $U(y_c) = c$. Expressing $S_y(\pm\infty)$ through the reflection and refraction coefficients one can find the relationship

$$\left(1 - |R|^2\right) \mathrm{Re}\, q_- - |T|^2 \mathrm{Re}\, q_+ = \pi \sum \left[\frac{\beta - U''(y)}{U'} |\psi|^2\right]_{y=y_c} . \qquad (5.64)$$

Here, as in Sect. 4.6, it is assumed that, for $\beta > 0$, the wave vector components are negative ($k_x < 0$, $q_\pm < 0$) so that the group velocity of the incident wave is directed towards the flow region and the phase velocity, outside this region. The critical layers for such waves exist only in the flows for which $U < 0$. Suppose that the conditions

$$U'(y) < 0, \quad U(-\infty) = 0, \quad U(+\infty) = -U_0 < 0 \qquad (5.65)$$

are fulfilled for the shear layer.

The left-hand side of (5.64) describes the energy gap at the Rossby wave reflection and refraction in a shear layer and the right-hand side, the value of absorption in the critical layers. One can see from (5.64) that resonance amplification is possible for the waves that are synchronous to the portions of the shear layer for which the condition

$$\left[\frac{d^2U}{dy^2}\right]_{y=y_c} < \beta \qquad (5.66)$$

is fulfilled.

Thus, resonant amplification is possible only for flows having a sufficiently "sharp" velocity profile. The amplification of the Rossby waves at reflection at the shear layer was first discussed by Lindzen and Tung (1978). The magnitude of resonant amplification was estimated numerically by Yamada and Goto (1983) and Yamada and Okamura (1984) for flows with the velocity profile of the form (5.15) and for jet profiles. Depending on the parameters of the flow and wave, the magnitude of relative amplification may amount to 10 when the wavelength is comparable with the characteristic size of the flow.

For smoothly inhomogeneous flows, one can neglect the profile curvature $d^2U/y^2 \ll \beta$ in (5.62) and use the WKB approximation. The boundary of the regions of propagation and nontransparency for the Rossby waves coincides with the critical layer y_c. In the neighborhood of this region, (5.62) degenerates to the reference equation (5.49). Its solution (5.50) describes the trapping and complete absorption of a monochromatic wave in the neighborhood of a singular point.

The inhomogeneity of flow velocity and stratification parameter M along the vertical coordinate may channel the wave field and form a waveguide for the Rossby waves. The eigenmodes of such a waveguide are calculated in the frame of (5.62) with appropriate boundary conditions. Similarly to IGW, a smooth horizontal inhomogeneity of the parameters of such a waveguide may give rise to the vertical focusing of the wave field near a certain level and to the trapping in the neighborhood of the critical layer at this level. The wave train tends asymptotically to the trapping point near which the wave energy is accumulated. These processes for the Rossby waves and the effects of viscosity, wave dispersion and nonlinearity in the neighborhood of the point of trapping were investigated by Gnevyshev and Shrira (1989).

Note in conclusion that the Kelvin–Helmholtz instability at the interface of two liquids of the same density but rotating with different frequencies was observed in laboratory experiments by M. V. Nezlin's group (Nezlin and Snezhkin, 1993). They also modelled Rossby vortices in rotating paraboloids. Such setups can be used for the experimental study of Rossby wave reflection at shear layers of inhomogeneous flows.

Chapter 6

Oscillatory Systems With Hydrodynamic Flows

The behavior in hydrodynamic flows of various oscillatory systems described by ordinary differential equations has always attracted researchers' attention, primarily, in terms of technical applications (e. g., Vol'mir, 1976, 1979). A continuous spatial spectrum of concentrated oscillators allows for radiation of a broad range of perturbations in a hydrodynamic flow. The integral effect of radiation reaction may lead in this case to qualitatively new phenomena. In this chapter we will consider some types of instabilities emerging as a result of the interaction of oscillatory systems with hydrodynamical flows. The mechanism of build-up is not related here to hydrodynamic flow instability: a free flow may be stable, destabilization occurring due to the feedback that is determined by the interaction with a concentrated oscillator.

6.1 Aerodynamic generation of sound

In aerodynamic acoustic generators (whistlers), the sound is excited due to the oscillatory motion of gaseous particles that is independent of the vibrations of the boundaries and is retained even if the boundaries are at rest (Zarembo and Krasil'nikov, 1966). The absence of moving parts in aerodynamic generators gives them certain advantages as compared to mechanical radiators. Besides everyday applications and signalling devices (Halton's whistle in steam boilers, police whistles, etc.), whistlers are broadly used in technologies (Chanaud, 1970; Shirokova, 1970). Finally, aerodynamic generators are commonly known to be employed in musical instruments. Flutes, organs, and some other instruments are different variants of whistles.

Note that aerodynamic generation is often a parasitic phenomenon. Spontaneous generation of a powerful whistle frequently accompanies operation of different devices and sometimes even leads to their destruction. For effective control of these sounds, one must get an insight into the process of their generation.

Two types of hydrodynamical processes leading to acoustic radiation may be distinguished. To the first group belong hydrodynamic oscillations emerging in the vortex flows where acoustic radiation is merely a side effect that does not influence the vortex motion. The main flow in this case may be described in the approximation of an incompressible fluid and the radiated sound may be calculated from the basic equations for an incompressible gas in the perturbation theory where the vortex motion is regarded as a given force. The motion of gaseous or fluid particles that leads to acoustic radiation may have different causes. One of the driving forces of such a motion is the turbulence accompanied by the vortex motion of particles. The vortex motion, in turn, may be non-steady-state, or translational "frozen" motion of vortices, or, finally, periodic motion as is the case with a flow around a cylinder at moderate Reynolds numbers (Batchelor, 1970).

Non-steady-state vortex motion gives rise to changes in pressure and, consequently, to sound excitation. It is a well-known Lighthill's mechanism of spontaneous radiation of sound by turbulence (Zarembo and Krasil'nikov, 1966).

The excitation of an acoustic resonator by the fronts moving in a nonstationary manner occurs analogously. The vortex motion produces pressure pulsations which, being a given external force, excite acoustic oscillations in a resonator. This phenomenon was analyzed by Blokhintsev (1981).

The vortex motion above the decelerating system may also give rise to surface waves in the system. Such a mechanism was proposed by Phillips (1977) as one of possible ways of wind wave generation (see Sect. 2.3).

Finally, periodic vortex shedding behind a bluff cylindrical body, that forms a periodic vortex wake and produces a periodic force acting on the cylinder may also lead to sound generation (aeolian tone). In particular, it is a reason why telegraph wires hum in windy weather.

The examples presented above are related to spontaneous acoustic radiation (radiation by the given vortices). The other type of aerodynamic acoustic generation is due to the sound–vortex feedback when the emerging sound gives rise to vortex oscillations. Such a feedback increases the efficiency of sound radiation.

A great number of various types of aerodynamic generators with acoustic feedback have been developed and investigated by now. Although some of them, for instance musical instruments, had been designed centuries ago (Mercer, 1951, 1953), the specific mechamism of feedback was studied relatively not long ago (Coltman, 1968; Andronov and Fabrikant, 1980). Below, a classical whistle, namely, an acoustic generator streamlined by an air flow will be considered. This type of whistles has important merits needed, in particular, for musical instruments, for instance, stability of the frequency specified by the resonator.

6.2 Linear theory of a whistle with a resonator

The mechanism of acoustic feedback was often discussed for flutes, organ pipes, and other whistles with resonators (Mercer, 1953). It was generally believed that the principal process here is an edge tone, i. e., the acoustic generation that occurs when a sonic stream crosses a wedge that is usually placed at the resonator neck (Brown, 1937). The resonator is then regarded only as an auxiliary element that amplifies the oscillations which otherwise do not depend on it. However, as was proved by Elder (1978), the basic mechanism of feedback in such whistles is determined by the elastic properties of the air column in the resonator.

Consider the simplest model of an organ pipe, which is actually an open end of an acoustic waveguide, with a plane flow moving above. Such a waveguide may be a neck of the Helmholtz resonator (Fig. 6.1a) or an end of an open acoustic resonator (Fig. 6.1b). Aerodynamic am-

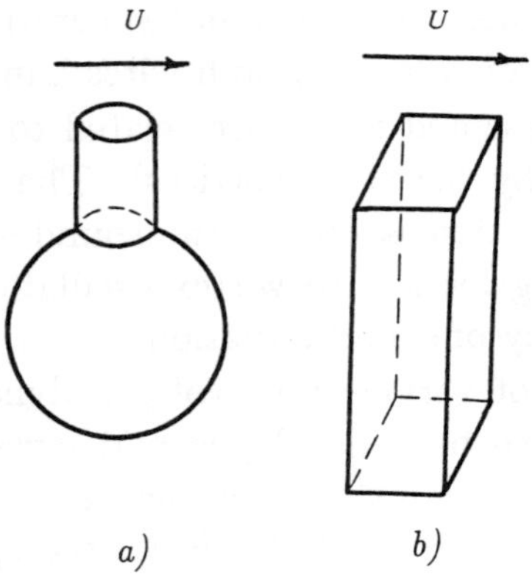

Figure 6.1: The simplest whistles:
(a) waveguide section, (b) Helmholtz resonator.

plification of oscillations in such a resonator with the air blown from the open end is determined by the effect of the flow on the reflection coefficient of the wave at the open end of the pipe. Assume that the characteristic size of the waveguide's cross-section is $a \ll c_s/\omega$ (where c_s is sound velocity and ω is frequency). Only the principal mode, for which the velocity and the shift of the particles of the medium are oriented along the pipe's axis, may propagate in such a waveguide. Then, the pressure and velocity oscillations in the waveguide far from the open end have the following form

$$\begin{aligned} p &= [\exp(i\omega y/c_s) + (R+\Delta R)\exp(-i\omega y/c_s)]\, p_0 \exp(-i\omega t), \\ v &= [\exp(i\omega y/c_s) - (R+\Delta R)\exp(-i\omega y/c_s)]\, v_0 \exp(-i\omega t). \end{aligned} \tag{6.1}$$

Here, R is the coefficient of the principal mode reflection at the open end of the waveguide. Neglecting the small correction ($\omega^2 s/c_s^2 \ll 1$ where s

is the area of the waveguide's cross-section) to the sound radiation from the open end (Landau and Lifshits, 1986), one can assume $R = -1$. The correction ΔR to the reflection coefficient is due to the presence of the flow. Note that the pressure node and the velocity antinode are at the open end ($y = 0$).

Calculation of the flow effect on the coefficient of wave reflection at the open end of a waveguide is, generally speaking, a rather difficult task (Veinstein, 1966). The correction ΔR, however, may be found approximately for the case when the effect of the flow is relatively weak, i. e., $|\Delta R| \ll 1$. This condition is fulfilled if the flow is far from the end of the pipe ($a \ll L$, see Fig. 6.1) or if the flow is a narrow jet. In these cases the excitation of the waveguide may be calculated in the first approximation by the perturbation reflected at the flow, as an external force, neglecting multiple re-reflections of the perturbations between the open end and the flow.

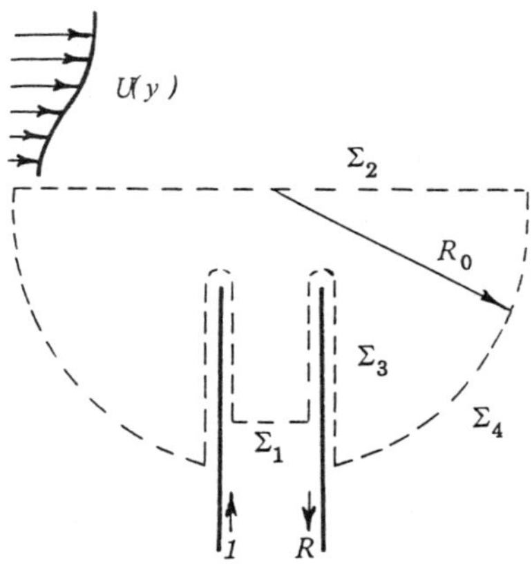

Figure 6.2: An open end of a waveguide near a shear flow. The surface of integration Σ is shown by the dashed line.

In order to relate the pressure perturbations inside and outside the waveguide consider the surface $\Sigma = \Sigma_1 + \Sigma_2 + \Sigma_3 + \Sigma_4$ bounding the closed region without a flow (see Fig. 6.2 where the surface Σ_3 borders on the waveguide walls and Σ_4 is a closing hemisphere of radius R_0).

Since the pressure perturbations in this region satisfy the Helmholtz equation

$$\Delta p + (\omega^2/c_s^2)p = 0, \tag{6.2}$$

the pressure $p^{(0)}$ in the absence of a flow and the perturbation $\tilde{p}$ introduced by the flow are related by the Green formula:

$$I\,(\Sigma) = \int\limits_{\Sigma} \left(p^{(0)} \frac{\partial \tilde{p}}{\partial n} - p \frac{\partial p^{(0)}}{\partial n} \right) dS = 0, \tag{6.3}$$

where n is the normal to the surface Σ. The boundary conditions $\partial p/\partial n = 0$ are met on rigid walls (the surface Σ_3). At the same time, as the radius R_0 of the closing surface Σ_4 is increased, the integral along this surface vanishes by virtue of the condition of the Sommerfeld radiation. Thus, the sum of the corresponding integrals with respect to the surfaces Σ_1 and Σ_2, i. e., $I(\Sigma_1) = I(\Sigma_2)$ may be taken to be zero.

Using the expression (6.1) for the sum $p^{(0)} + \tilde{p}$ and the expression $p_0 = \rho_0 c_s v_0$ relating the pressure and velocity amplitudes of the principal wave, one can find

$$I\,(\Sigma_1) = \int\limits_{\Sigma_1} \left(p^{(0)} \frac{\partial \tilde{p}}{\partial n} - \tilde{p} \frac{\partial p^{(0)}}{\partial n} \right)_{y=L-R_0} dS = -2i\omega S_0 \rho_0^2 c_s \Delta R v_0^2, \tag{6.4}$$

where ρ_0 is density and S is the area of the waveguide's cross-section. Passing to the Fourier transforms for $y > 0$:

$$\begin{aligned} p^{(0)} &= \int H_{\mathrm{k}}^{(0)} \exp\left[\kappa(k)y + i(\mathbf{ks})\right] d\mathbf{k}, \\ \tilde{p} &= \int \left[\tilde{G}_{\mathrm{k}} \exp\left(-\kappa y\right) + \tilde{H}_{\mathrm{k}} \exp\left(\kappa y\right)\right] \exp i(\mathbf{ks})\, d\mathbf{k}, \end{aligned} \tag{6.5}$$

where $\kappa^2 = k^2 - \omega^2/c_s^2$ ($\mathrm{Re}\,\kappa > 0$), $\mathbf{k} = (k_x, k_z)$, and $\mathbf{s} = (x, z)$, one finds

$$I(\Sigma_2) = \int\limits_{\Sigma_2} \left(p^{(0)} \frac{\partial \tilde{p}}{\partial n} - \tilde{p} \frac{\partial p^{(0)}}{\partial n} \right)_{\mathrm{y=L}} dS = 8\pi^2 \int \kappa H_{-\mathrm{k}}^{(0)} G_{\mathrm{k}} e^{-2\kappa L} d\mathbf{k}. \tag{6.6}$$

The amplitude $H_k = (H_k^{(0)} + \tilde{H}_k)e^{-\kappa L}$ of the pressure perturbation incident on the flow and the amplitude $G_k = \tilde{G}_k e^{\kappa L}$ of the reflected perturbation are related by the reflection coefficient $R(\omega, k) = G_k/H_k$ the magnitude of which was found by Andronov and Fabrikant (1980) for different flows. The fluid motion may be considered to be incompressible at distances $L \ll c_s/\omega$, which simplifies further calculations. Assuming the perturbation reflected at the flow to be small, one can set to the first approximation

$$\tilde{G}_k = (H_k^{(0)} + \tilde{H}_k)e^{-2\kappa L} = RH_k^{(0)}e^{-2\kappa L}. \tag{6.7}$$

Finally, by expressing, from the linearized Euler equation

$$\frac{\partial v}{\partial t} + \frac{1}{\rho_0}\frac{\partial p}{\partial y} = 0$$

in the flowless region, the amplitude of perturbation of pressure $H_k^{(0)}$ through the amplitude of the y-component of the velocity perturbation $v_k^{(0)}$ $(y \neq 0)$:

$$H_k^{(0)} = \frac{-i\omega\rho_0}{\kappa}v_k^{(0)}(y = 0) \tag{6.8}$$

and equating (6.4) and (6.6) one gets

$$\Delta R = -\frac{4\pi i\omega}{cS_0}\int \frac{R(\omega, k)W}{\kappa(k)}e^{-2\kappa(\omega,k)L}d^2\mathbf{k}. \tag{6.9}$$

Here, $W = v_k^{(0)}v_{-k}^{(0)}/4v_0^2 = |v_k^{(0)}/2v_0|^2$ because, neglecting viscosity and compressibility, the relation $v_k^{(0)}/2v_0$ is the Fourier transform of the real function, i. e., of the y-component of the velocity of the potential flow of a perfect incompressible fluid that occurs as a result of potential outflowing from the end of the pipe inside which the velocity is equal to unity (Fig. 6.3).

The velocity distribution at incompressible outflow from the pipe is very complicated (see Lamb, 1932); the magnitude of W can be easily found only in two cases:

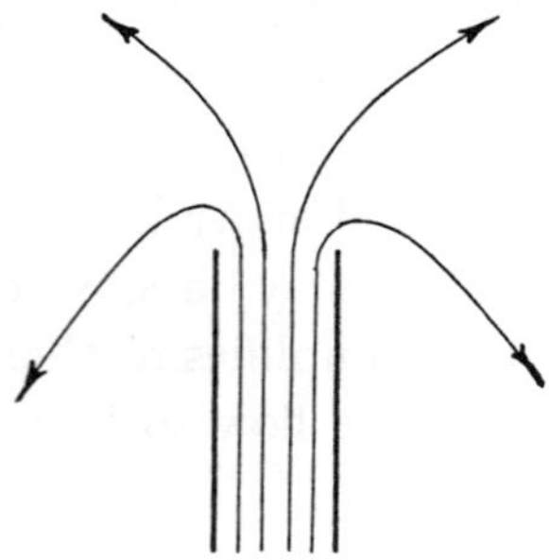

Figure 6.3: Streamlines for the potential flow from an open end of a tube.

i) when the longitudinal (a) and transversal to the flow (b) sizes of the pipe's cross-section are small enough ($a, b \ll |r|$, where the three-dimensional radius-vector $\mathbf{r} = (x, y, z)$) and the flow can be considered to be isotropic:

$$\mathbf{v} = \frac{v_0 S_0 \mathbf{r}}{4\pi |\mathbf{r}|^3}, \tag{6.10}$$

ii) and when $a \ll |\mathbf{r}| \ll b$ so that the flow is two-dimensional and depends on a plane radius-vector $\boldsymbol{\rho} = (x, y)$:

$$\mathbf{v} = \frac{v_0 a \boldsymbol{\rho}}{2\pi |\boldsymbol{\rho}|^2}. \tag{6.11}$$

Let us now choose the integration path in (6.9). The choice is made by considering the initial value problem, i. e., by "switching in" oscillations in the resonator. Let us employ the Laplace method and consider the complex values of ω far from the upper halfplane, when all the poles in the integrand in (6.9) lie in the upper halfplane of the complex k_x. Then, the integration can be performed along the real axis of k_x.

The expression obtained for ΔR can be extended to the real ω. To this end, one must deform the integration path in the complex plane of k_x so that the poles remain above them. As a result one obtains a contour that consists of the real axis k_x supplemented by the loops tracing below all the poles of the integrand in the complex plane of

k_x (Fig. 6.4). The branching of the function $\kappa(k)$ ($\mathrm{Re}\,\kappa > 0$) at the points $k_x = \pm\eta$, where $\eta = \sqrt{\omega^2/c_s^2 - k_z^2}$, does not affect the choice of the integration path because the loops in the complex plane do not intersect the cuts.

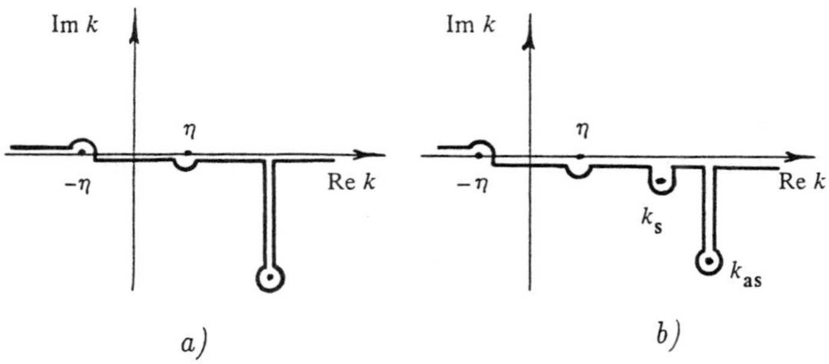

Figure 6.4: The integration path in the expression for the reflection coefficient Δr for a vortex sheet (a) and for a jet (b).

These arguments are completely analogous to the ones used in Landau's rule of path-tracing in the theory of plasma oscillations as well as in Lin's rule of path-tracing in the theory of hydrodynamic instability (see Sect. 2.2). The only difference is that in the two above cases the shift of the integration path was insignificant and corresponded to the rule of the pole path-tracing along the real axis. In our case, the shift of the integration path may be pronounced because it is necessary to trace the poles that are far from the real axis. Such an integration path in a complex plane was used in the investigations concerned with source radiation near a vortex sheet (Jones and Morgan, 1972; Mironov, 1975).

The same integration path may, evidently, also be obtained for the real ω if one takes into account the boundary conditions at infinity, $x \to \pm\infty$. Because the flow moves from the region $x = -\infty$ to the region $x = +\infty$, the perturbations decrease as $x \to -\infty$, but may grow as $x \to +\infty$ due to the instability of the intrinsic oscillations of the flow. The functions of this form are represented by the integrals in the complex plane k_x along the contours of the type specified above. The poles of $R(\omega, k)$ traced by the loops in Fig. 6.4 correspond to unstable intrinsic oscillations of the flow.

The correction ΔR to the coefficient of the principal wave reflection at the open end of the waveguide, that was obtained above to the first approximation, determines the amplification of the reflected wave, given

$$\mathrm{Re}\,(R\Delta R^*) \simeq -\mathrm{Re}\,(\Delta R) > 0, \quad \text{i. e.,} \quad \mathrm{Re}\,(\Delta R) < 0.$$

This amplification may give rise to the self-excitation of the resonator part of which is a waveguide. For example, the growth rate of build-up in an organ pipe (Fig. 6.1a) in the absence of dissipation is equal to

$$\gamma = -\frac{\omega \mathrm{Re}\,(\Delta R)}{2\pi n} = -\frac{2\omega^2}{ncS_0}\mathrm{Im}\int \frac{R(\omega,k)W}{\kappa(k)} e^{-2\kappa(\omega,k)L}\,d\mathbf{k}, \qquad (6.12)$$

where ω is the frequency of the intrinsic oscillations of the resonator and n is the number of the wavelengths that fit into the pipe.

Consider now the quantity $\mathrm{Re}\,(\Delta R)$ for different shear flows. The reflection coefficient $R(\omega, k)$ has no poles and $\mathrm{Im}\,R = 0$ for a vortex sheet with velocity discontinuity U_0 at real k_x (see (5.2) at $c_s = \infty$). The magnitude of amplification is determined only by one pole in (6.9) $k_x = k_1 = \omega(1-i)/U_0$ in the upper half-plane of complex k. By using the expression (6.10) for a thin pipe located far from the vortex sheet one finds

$$\mathrm{Re}\,(\Delta R) = \frac{\omega S_0}{2c_s L}\sqrt{\frac{\omega L}{\pi U_0\sqrt{2}}}\exp\left(-\frac{2\omega L}{U_0}\right)\sin\left(\frac{2\omega L}{U_0}+\frac{\pi}{8}\right). \qquad (6.13)$$

For the plane waveguide (the two-dimensional flow (6.11)) one has analogously

$$\mathrm{Re}\,(\Delta R) = \frac{\omega a}{c_s\sqrt{2}}\exp\left(-\frac{2\omega L}{U_0}\right)\sin\left(\frac{2\omega L}{U_0}+\frac{\pi}{4}\right). \qquad (6.14)$$

These expressions hold only if $\omega L/U_0 \ll 1$. Otherwise, the reflection at the flow cannot be considered to be small and the approximation (6.7) is not valid.

One can see from (6.13) and (6.14) that the sign and magnitude of amplification of oscillations in the pipe depend on the values of the parameter $\omega L/U$. The latter determines the development at length L (L

is the length of the path of flow perturbations past the resonator) of the Kelvin–Helmholtz instability of the vortex sheet, that is excited by the open end of the pipe. The fact that ΔR is an alternating function is attributed to the dependence of the phase of the unstable perturbation on the distance to the vortex sheet because the wavenumber of the surface wave has an imaginary part on the vortex sheet. This dependence of the amplification of oscillations in an acoustic resonator on the distance L to the flow, i. e., the alternating growth rate and its rapid decrease with increasing L, was observed in experiments on the excitation of acoustic oscillations by a shear layer in cavities (Elder, 1978).

It is worthy of notice that the allowance made for weak "smearing" of monovelocity flows (a vortex sheet, as well as a jet that will be considered below) gives a small addition $\operatorname{Im} R(\omega, k)$ at real ω and k and slightly displaces the pole k_1. This leads only to minor changes of the magnitude of amplification. The flow as a whole rather than the resonant particles (cf. Sect. 2.2) takes part in amplification of the oscillations in the resonator.

Consider another plane-parallel flow which has small ΔR and the approximate expression (6.7) is now valid. It is a thin jet of length l with a piecewise constant velocity profile

$$U = \begin{cases} U_0 & \text{for} \quad |y - L| < l/2, \\ 0 & \text{for} \quad |y - L| > l/2. \end{cases} \tag{6.15}$$

The reflection coefficient of pressure perturbations for this jet in the approximation of an incompressible fluid can be written as

$$R(\omega, k) = \frac{[(\omega - kU_0)^4 - \omega^4] \tanh(kl) \exp(-2kL)}{[(\omega - kU_0)^4 + \omega^4] \tanh(kl) + 2\omega^2(\omega - kU_0)^2} \tag{6.16}$$

and has two poles in the upper half-plane k_x. For $\omega l/U_0 \ll 1$ these poles

$$k_{\mathrm{s}} = \left(1 + i\sqrt{\frac{\omega l}{2U_0}}\right), \qquad k_{\mathrm{as}} = \sqrt[3]{\frac{2U_0}{\omega l}} \frac{\omega(\sqrt{3} + i)}{2U_0} \tag{6.17}$$

correspond to the symmetric (pulsed) and antisymmetric (flexural, see Fig. 6.5) eigenmodes of the jet (Betchov and Criminale, 1967).

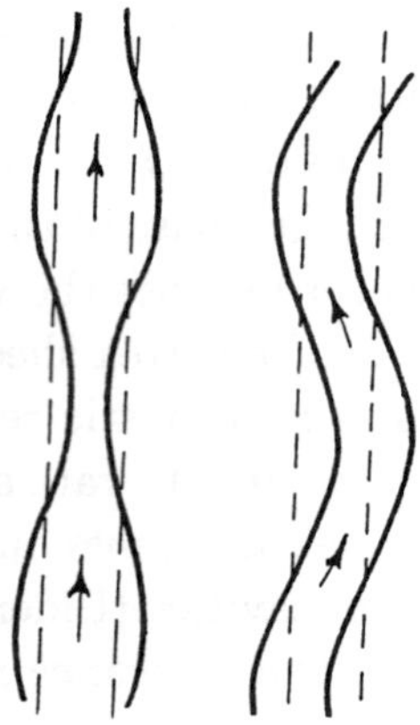

Figure 6.5: Eigenmodes of oscillations of a thin jet: (a) pulsed ("sausage") mode, (b) flexural ("lenticular") mode.

If the quantity $\omega L/U_0 \gg l/L$, then the pole k_{as} makes an exponentially small contribution to ΔR so that the reflection coefficient is determined only by the symmetric mode of the thin jet. For a two-dimensional flow (when the transverse size of the pipe opening $b \gg L, a, U_0/\omega$), to the first approximation with respect to $\omega l/U_0$, one has

$$\begin{aligned}\mathrm{Re}\,(\Delta R) &= -\frac{8\pi\omega l}{ac_s}\mathrm{Im}\int W\frac{\omega^2\exp(-2k_xL)}{2(\omega-k_xU_0)^2}\frac{dk_x}{k_x}\\ &= -\frac{4\pi^2\omega^3 l}{ac_sU_0^2}\frac{\partial}{\partial k}\left(We^{-2kL}\right)_{k=\omega/U_0}.\end{aligned} \tag{6.18}$$

Thus, the harmonics of perturbations in a thin jet that are synchronous to the flow velocity contribute to the amplification of the reflected wave. It is essential how the spectrum of velocity oscillations behaves at the open end at a synchronous wavenumber $k_0 = \omega/U_0$. If the spectrum of W grows with increasing k, then the jet may give up its energy to the oscillations in the waveguide and thus amplify them.

The latter circumstance has a simple explanation: The particles of the jet that are in the perturbation phase, in which they decrease their velocity and give up energy, dominate over the particles in the accelerating phase because, as the velocity is decreased, the resonance wavenumber $k = \omega/U$ and the perturbation spectrum grow. In other words, the jet interacts stronger with the lagging behind perturbation harmonics having a broader power spectrum.

Apparently, the derivatives of the perturbation spectrum here play the same role as the derivatives of the velocity profile $U(y)$ at the monochromatic wave interaction with a shear flow that was considered in Chap. 2.

Analysis of (6.18) for a far jet ($a \ll L$) gives

$$\mathrm{Re}\,(\Delta R) = \frac{\omega a}{2c_s}\frac{\omega^2 lL}{U_0^2}\exp\left(-\frac{2\omega L}{U_0}\right) > 0. \tag{6.19}$$

The thin jet, in this case, attenuates the oscillations in the pipe.

When close to the open end, the jet amplifies the oscillations in the regions of the positive derivative $(\partial W/\partial k)_{k=\omega/U_0}$. The size of the waveguide neck a is the characteristic spatial scale of the oscillatory amplitude variations along the x-axis. Therefore, $\Theta = \omega a/U$ is the parameter determining the behavior of the W spectum. In particular, taking the simplest model for the amplitude of velocity oscillations

$$v(y=0) = \begin{cases} 2v_0 & \text{for} \quad |x| < a/2; \\ 0 & \text{for} \quad |x| > a/2, \end{cases} \tag{6.20}$$

one has the spectrum of

$$W = \frac{\sin^2(ka/2)}{(\pi k)^2}$$

so that a value $(\partial W/\partial k)_{k=\omega/U_0}$ and, hence, $\mathrm{Re}\,(\Delta R)$ will oscillate with the variation of Θ, changing the sign as Θ is increased by the value of order π. This result is in a good qualitative agreement with the experiments carried out by Coltman (1968) who measured the impedance of the streamlined open end of the waveguide.

6.3 Acoustic analogs of electron micro-wave oscillators

The mechanism of amplification of the acoustic oscillations due to their interaction with the eigenmodes of a shear flow in some types of whistles is very similar to the mechanism of generation of electromagnetic oscillations in electon microwave oscillators. The analogy of acoustic and electron devices may be useful for the understanding of the general principles of excitation of oscillations and, probably, for designing new types of amplifiers and generators. Below we will describe the operation of some electron oscillators and the corresponding aerodynamic sound generators and will, thus, continue the plasma–hydrodynamic analogy that was presentd in Sect. 2.2 (Andronov and Fabrikant, 1979).

6.3.1 Travelling wave tube

If the plasma beam has a single velocity, so that the function of the particle velocity distribution $f(v_x) \sim \delta(v_x - v_0)$ (a "cold" beam in which there is no thermal spreading of particle velocities), the value of conductivity $\operatorname{Re}\sigma \to \infty$ considered in Sect. 2.2 at $\omega/k \to v_0$ and the relation for the instability growth rate $\gamma \sim \operatorname{Re}\sigma$ are no longer valid because they hold only for $\operatorname{Re}\sigma \ll \operatorname{Im}\sigma$.

In this case, however, one can find the growth rate directly from the dispersion equation for longitudinal waves in a plasma with a beam. This equation for the longitudinal dielectric permittivity

$$\varepsilon_l = 1 - \frac{\omega_p^2}{\omega^2} - \frac{\omega_b^2}{(\omega - kv_0)^2} = 0, \tag{6.21}$$

where ω_p is the Langmuir frequency of a plasma at rest and ω_b is the Langmuir frequency of the beam (Mikhaĭlovskiĭ, 1974), takes on the form

$$(\omega^2 - \omega_p^2)(\omega - kv_0)^2 = \omega_b^2\omega^2. \tag{6.22}$$

The solutions of this equation, as $\omega \to 0$, are the intersecting dispersion curves for waves in a plasma and in a beam. The degeneration at the intersection point is removed at $\omega_b \neq 0$ and the instability emerges in the neighborhood of this point $k = \omega/v_0$ (see Fig. 1.13 which depicts the dispersion curves for the interaction of two waves with energies of opposite signs). Such a beam instability is referred to as "hydrodynamic instability" because the beam as a whole takes part in production of the growth rate (in contrast to the kinetic instability of a warm beam, like the one shown in Fig. 2.2, in which only resonance particles with the velocity $v_x = \omega/k_x$ interact with the wave).

Such a hydrodynamic instability is used in travelling wave tubes (TWT) for the amplification of microwave oscillations. These electron devices typically have a slowing down system that "carries" an electromagnetic surface wave, and an electron beam that is transmitted along this system (Pierce, 1974). The instability occurs when the dispersion curve of the surface wave intersects the dispersion curve of the negative energy Langmuir wave of the beam.

Consider now the acoustic analog of a TWT, namely, a shear hydrodynamic flow above an acoustic slowing down system. An example of such a system is a well-known "comb" that is a chain of acoustic resonators (Fig. 6.6). The fields at the open ends are related so that

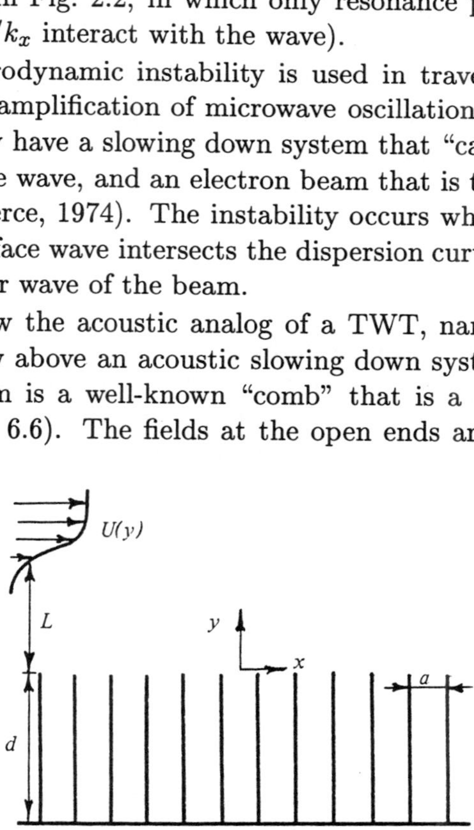

Figure 6.6: Acoustic analog of electronic travelling wave tube.

a surface wave with the phase velocity $\omega/k \ll c$ may travel along this "comb" (Gyan Dyn Hua, 1961). The impedance of a comb, of the depth d, calculated for the surface $y = 0$ bordering on the open ends of the resonators has the form

$$Z \equiv \left[\frac{p(\omega, k)}{v(\omega, k)}\right]_{\mathrm{y}=0} = ic_{\mathrm{s}}\rho \operatorname{ctg}\frac{\omega d}{c_{\mathrm{s}}}. \tag{6.23}$$

Experimental investigations of the flow effect on the surface wave propagation along a periodic structure (Meyer, 1963) demostrated that the wave may be amplified by subsonic jets. The analogy of this hydrodynamical model to the electron TWT appears to be quite obvious. However, the mechanism of wave amplification in an "acoustic TWT" is different from that in its electron prototype.

The dispersion equation for waves in the "comb"–flow system may be found from the solution of the Helmholtz equation (6.2) in the region between the "comb" and the flow:

$$p = Ge^{\kappa(y-L)} + He^{-\kappa(y-L)}, \tag{6.24}$$

where G and H are some constants.

By using the expression for the reflection coefficient $R(\omega, k) = G/H$ of the pressure perturbation amplitudes and the impedance boundary condition (6.23) one can obtain the dispersion equation

$$\omega + \frac{i\kappa}{\rho_0}Z(\omega, k) = \left[\omega - \frac{i\kappa}{\rho_0}Z(\omega, k)\right] R(\omega, k)e^{-2\kappa L}. \tag{6.25}$$

The solutions of this equation, for $kL \ll 1$, are two weakly coupled wave modes: the surface wave on the "comb" with the dispersion equation

$$\omega + \frac{i\kappa}{\rho_0}Z(\omega, k) = 0$$

and the normal wave of the flow the dispersion equation for which is determined by the pole of the reflection coefficient: $R^{-1} = 0$. Free shear flows (vortex sheet, shear layer, jet) are unstable, as a rule, so that the dispersion equations for them have no real roots (Betchov and Criminale, 1967). Therefore, the dispersion curves of the surface wave and of the intrinsic oscillations of the flow do not intersect in this case and thus no strong wave coupling occurs. The weak coupling of the two waves only slightly changes the intrinsic frequency of the surface wave and does not give rise to instability. For instance, the substitutions of (5.2) for $R(\omega, k)$ and of (6.23) for the comb's impedance into the dispersion equation (6.25) give for the vortex sheet

$$\frac{(\omega - kU_0)^2 + \omega^2}{(\omega - kU_0)^2 - \omega^2} = -\frac{\omega + c_s\kappa \operatorname{ctg}(\omega d/c_s)}{\omega - c_s\kappa \operatorname{ctg}(\omega d/c_s} e^{-2kL}. \tag{6.26}$$

One can show that the roots of (6.26) corresponding to the surface waves of the "comb" remain real at $L \neq 0$.

Thus, the "monovelocity" hydrodynamic flows cannot amplify (or attenuate) the slowed down monochromatic waves. Here, only the Miles amplification mechanism is possible which corresponds to the mechanism of the kinetic instability of the plasma beam. This mechanism was considered in Sect. 2.3 for wind waves for which the surface of a heavy liquid surves as a "slowing down" structure. The Miles mechanism holds only for the "smeared" velocity profile where $d^2U/dy^2 \neq 0$. The quantity $\operatorname{Im} R(\omega, k)$ may then be nonzero, which provides the energy exchange between the wave and the flow. The instability growth rate of the surface wave in this case is

$$\gamma = \frac{2\omega}{1 + \partial(ikZ/\rho_0)/\partial\omega} \operatorname{Im} R(\omega, k). \tag{6.27}$$

So, unlike TWT, only a small part of the beam, i. e., the resonant particles concentrated at the "slopes" of the profile $U(y)$ interact with the wave. In order to produce the conditions under which the flow as a whole takes part in amplification, one needs to consider the interaction of the flow with the systems whose spatio-temporal spectrum is not monochromatic but, instead, has components in the region corresponding to the intrinsic oscillations of the flow. This situation occurs in whistles.

6.3.2 Monotron

Consider quasistatic electronic amplifiers in which the spatial field distribution is finite. An example of such a system is an electron gap, i. e., a parallel-plate capacitor penetrated by the electron beam with applied harmonic voltage $\tilde{V}$ (Fig. 6.7, left). The electric field of such a gap is harmonic in time but has a continuous spatial spectrum (Fig. 6.7, right). Clearly, for determination of the gap conductivity, σ_c, one needs to integrate the work done by the individual spectral components of the field. Consequently,

$$\mathrm{Re}\,\sigma_c \sim \int \mathrm{Re}\,\sigma(\omega, k)|E_k|^2\,dk. \tag{6.28}$$

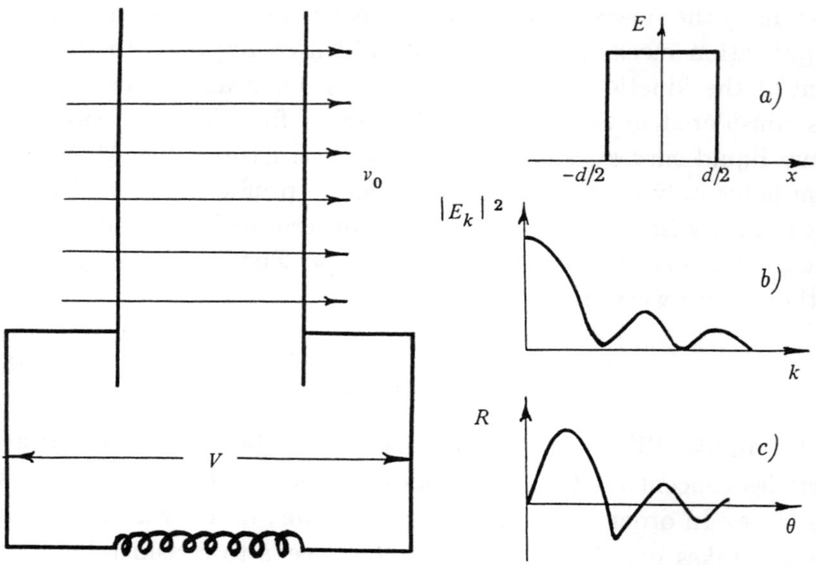

Figure 6.7: Scheme of monotron (left) and its characteristics (right):
(a) spatial distribution of the electric field $E(x)$,
(b) energy spectrum of the field $|E_k|^2$,
(c) resistance R as a function of the angle of flight θ.

Because, for the phase velocities close to the beam velocity ($c \approx v_0$), $\mathrm{Re}\,\sigma(\omega, k)$ has an alternating sign (Fig. 2.2), it is apparent that the sign of $\mathrm{Re}\,\sigma_c$ depends on which components prevail in the spectrum $|E_k|^2$:

the ones with $c > v_0$ or with $c < v_0$, i. e., it, actually, depends on the derivative $(\partial|E_k|^2/\partial k)_{k=\omega/v_0}$. This can readily be found by direct calculations by considering a monovelocity beam for which

$$\sigma(\omega, k) = -\frac{\omega_b^2}{v_0}\frac{\partial}{\partial k}\delta(\omega - kv_0). \tag{6.29}$$

The result is a well-known expression for the monotron resistance

$$R_m = \frac{4\pi d}{\omega S}\frac{\omega_b^2}{\omega^2}\frac{\theta\sin\theta + 2\cos\theta - 2}{\theta}, \tag{6.30}$$

where S is the area of the monotron's plates and $\theta = \omega d/v_0$ is the "angle of flight", i. e., the parameter on which the sign of the resistance depends (see Fig. 6.7).

The impedance of the electron gap, $Z_m = R_m + iX_m$, may be found, for example, by the Fourier method. This method is convenient for hydrodynamical problems, although such a calculation technique is not traditional for electronics (cf. Gaponov, 1960; Andronov and Chugunov, 1975). The calculations of the potential difference between the plates from the Poisson equation with the charge concentrated at the plates and with the dielectric permittivity of the beam

$$\varepsilon_l = 1 - \frac{\omega_b^2}{(\omega - kv_0)^2} \tag{6.31}$$

give (Feix, 1964; Andronov and Chugunov, 1975)

$$Z_m = \frac{8}{i\omega S}\int \frac{\sin^2(kd/2)}{k^2\varepsilon_l(\omega, k)}\,dk. \tag{6.32}$$

The monotron resistance $R_m = \operatorname{Re} Z_m$ is determined by the poles of the integrand at $\varepsilon_l = 0$ on the real axis k at the points

$$k_\pm = \frac{\omega \pm \omega_b}{v_0}. \tag{6.33}$$

The rule of path-tracing (in the lower part of the complex plane k) of these poles may be found similarly to the corresponding rule for a whistle (see Sect. 6.2), i. e., by solving the initial value problem. An alternative method of calculations is to take into account small dissipation in the dielectric permittivity ε. As $\omega_b \to 0$, both the poles merge into one second-order pole, which gives the expression (6.30) for R_m.

Consider now the monotron in a plasma with a beam (Andronov and Fabrikant, 1979; Petelin, 1985). The beam in a plasma is an unstable system even if the influence of the gap is not taken into account. Let us assume, however, that an unperturbed beam enters the monotron and the beam instability is developing only starting from the instant the beam enters the gap and is determined by the field perturbations induced by the charge on the plates. The dielectric permittivity of the plasma beam (6.21) at $\omega < \omega_p$ has no zeros on the real axis k. However, the integration path in (6.32) must be taken so that it should embrace from below all the poles in the complex plane k. Similarly to the theory of a whistle, this path may be found from the solution of the initial value problem by the Laplace method. With the path-tracing rule taken into account, one can find from (6.32) the resistance R that is found by subtracting from the pole $k_+ = \omega/v_0 - i\omega_b/v_0\sqrt{|\varepsilon_0|}$ (where $\varepsilon_0 = 1 - \omega_p^2/\omega^2$) that corresponds to the waves increasing downstream. The poles $k_\pm$ move towards the real axis, as ω_b is decreased, so that only one second-order pole remains on the real axis and the resistance of the gap differs from the value R_m in (6.30) only by the factor $|\varepsilon_0|^{-2}$.

This example illustrates the application of the path-tracing rule in the integration with respect to the wavenumber spectrum when there are no "doubtful" points. Apparently, the integration path that traces the poles in the lower halfplane of the complex k cannot be found from the contour along the real axis by taking small dissipation into account. However, it can be obtained by considering a boundary value problem for the gap assuming that the beam entering the gap is unperturbed.

Note that the development of the beam instability corresponding to the pole k_+ plays the same role as in the classical "electron bunching–intrinsic beam oscillations" monotron. Thus, for $R_m < 0$ to occur in the gap, the presence of intrinsic oscillations of the beam is essential, while it is of no significance whether they are damping or increasing.

It is clear from what was said above that the mechanisms of amplification of oscillations are very similar in a whistle with a resonator and in a monotron. In both the cases, the magnitude and the sign of amplification of the oscillations in the resonator depend on the development of wave flow perturbations at the length of its drift past the resonator, namely, by the variation of the phase of these perturbations (the "angle of flight") $\theta = \omega l/U$. The distance L to the flow, if it is far from the resonator neck, or the size of the neck a, if the flow is close to the neck (like is the case of the thin jet considered above), is the characteristic size l at which the effective interaction of the acoustic oscillations with the flow occurs. The energy exchange between the flow and the oscillations depends on the derivative of the spectrum of oscillations at resonant harmonics for both, a whistle with a thin jet and a monotron.

Unlike hydrodynamic flows, the oscillations of a free electron beam are stable. However, the investigation carried out in Sect. 6.2 for the interaction of a thin jet with oscillations in an acoustic waveguide shows that here, like in an electronic microwave amplifier, of primary importance for the excitation of oscillations is the presence of the intrinsic oscillations of the flow and its inertial properties rather than the flow instability. The latter circumstance was noted by Konstantinov (1974). For a thin jet, the development of the perturbations of a weakly unstable symmetric mode at the length of the drift past the open end of the waveguide plays the role that is similar to the electron bunching in microwave devices. The external impulse (from the resonator side) produces velocity perturbations in the jet which are transformed to pressure perturbations in the course of the development of the jet's eigenmode. The pressure perturbations, again, excite the resonator, thus closing the feedback loop.

6.3.3 Klystron

A well known variety of electron microwave oscillators is a klystron that includes two electron gaps penetrated by a common beam of electrons (Gaponov, 1960) (Fig. 6.8a). The field structure, spatial spectrum, and resistance of such a system are presented in Fig. 6.9.

The klystron may be compared to the whistle depicted in Fig. 6.8b.

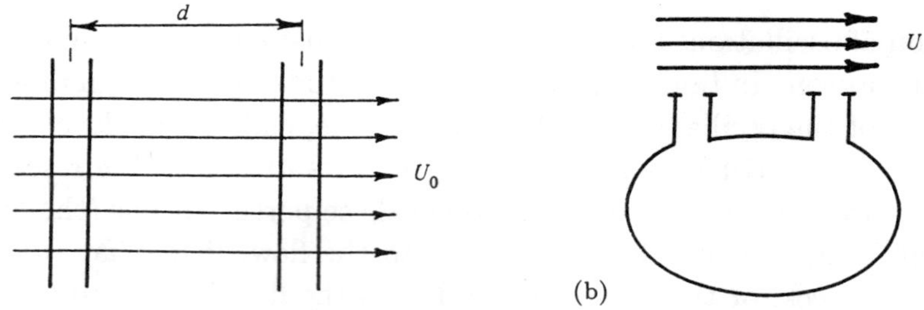

Figure 6.8: Klystron: (a) electromagnetic, (b) acoustic.

It is the Helmholtz resonator with two necks streamlined by a common flow. The perturbations produced in the shear flow by the first neck are amplified by the flow and induce the alternating pressure in the second neck.

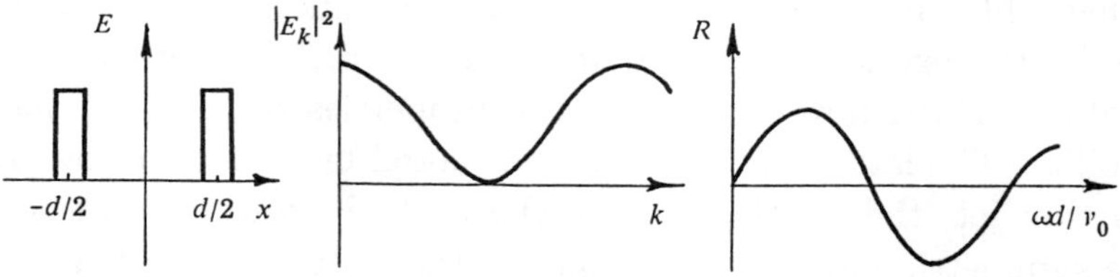

Figure 6.9: Klystron characteristics (see the notation in Fig. 6.7).

Assume for simplicity that the sizes and shapes of the two necks are identical. Then, the velocity oscillations in the necks are in phase and have equal amplitudes. Let us make the calculations like the ones described in Sect. 6.2 and take the surface Σ_1 that consists of two parts: the cross-sections of the two necks. Then, the corresponding integral in (6.3) will be equal to the doubled expression (6.4). Whereas the integral with respect to the surface Σ_2 retains the form (6.6) so that, instead of (6.9), the correction to the coefficient of wave reflection in each neck is written as

$$\Delta R = -\frac{8\pi i\omega}{c_s S_0}\int \frac{R(\omega,k)W}{\kappa(k)} e^{-2\kappa(\omega,k)L}\, dk, \tag{6.34}$$

where W is the spatial power spectrum of the oscillatory y-component of the velocity of the flow that is formed as the fluid is outflowing from two parallel tubes inside which the velocity is equal to unity.

Consider the amplification of oscillations in this resonator by a flow under the assumption like the one made above, and consider the flow to be two-dimensional. The latter is valid when both necks have rectangular cross-sections for which the size transverse to the flow is $b \gg a, l, d, U_0/\omega$ where d is the distance between the two necks. Then, instead of (6.18), for the thin jet located rather close to the resonator ($L \ll U_0/\omega$) one has

$$\operatorname{Re} \Delta R = -\frac{2\pi^2 \omega^3 l}{a c_s U_0^2} \frac{\partial W}{\partial k} \Big|_{k=\omega/U_0} . \qquad (6.35)$$

The spectrum of W is readily calculated for narrow necks when $a \ll l, U/\omega$. In this case, one can assume

$$v = a\left[\delta(x) + \delta(x-d)\right], \quad W = 4a^2 \cos^2(kd/2). \qquad (6.36)$$

The quantity $(\partial W/\partial k)_{k=\omega/U_0}$ and the growth rate of the oscillations in the resonator are specified by the change of the phase of jet's intrinsic oscillations at the drift length between the necks, i. e., by the parameter $\theta = \omega d/U_0$, and vary with the increase of this parameter. The same angular dependence is typical of a klystron where the amplification of electromagnetic oscillations is determined by the development of the eigenmodes of the electron beam (electron bunching) at the length of the drift between the resonators. Perhaps, the whistle with a "double" resonator will have the merits of a klystron (as compared to a monotron which is an analog of an ordinary whistle); in particular, it may have a high efficiency (Gaponov, 1960).

6.4 Radiative instability of streamlined oscillators in a moving fluid

Consider an example illustrating still another mechanism of build-up of a concentrated oscillator in a hydrodynamic flow. This mechanism involves the radiation of waves into a medium moving with a "superwave" velocity relative to the oscillator, i. e., with the velocity U that is higher than the phase velocity c_p of the waves at the oscillator frequency.

The radiative instability of moving oscillators is well-known in electrodynamics (Ginzburg, 1975). This effect in hydrodynamics was investigated by Gaponov–Grekhov *et al.* (1983) on an example of a simple model of mechanical oscillator — a ball on a spring stretched with constant velocity U parallel to the interface of two liquids of different densities at distance h from the interface (Fig. 6.10). The moving body

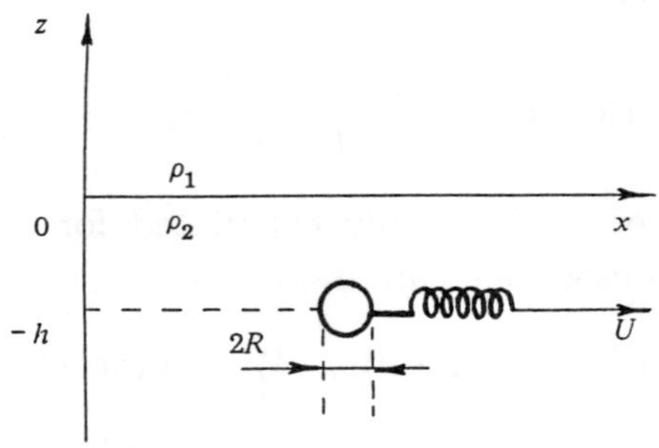

Figure 6.10: Mechanical generator moving above the interface between two liquids of different densities.

excites waves on the surface of the liquid interface and experiences the force of radiative resistance. This force consists of two components: the constant one, F_0, balanced by the tension of the spring and the oscillatory component $\tilde{F}$. The equation for the ball's oscillations is written in the form

$$\frac{d^2\xi}{dt^2} + \omega_0^2\xi = \frac{\tilde{F}}{m}, \tag{6.37}$$

where ξ is the oscillating part of the displacement of the ball, ω_0 is the intrinsic frequency of the oscillator, and m is the mass of the ball, the added mass included.

The radiative force may be calculated from energy considerations in a linear approximation, when the oscillatory frequency is assumed to be close to ω_0, by equating the average work done by this force per unit time, $A = \langle \tilde{F} d\xi/dt \rangle$, to the radiated power. The latter may be found from the problem of the radiation of a small sphere near the interface. The solution to this problem (Gaponov–Grekhov *et al.*, 1983) enables one to find both the constant component of the force of radiative resistance, F_0, that arises due to the "Cherenkov" radiation of ship waves and the oscillatory component which consists of two parts: $\tilde{F} = \tilde{F}_+ + \tilde{F}_-$, where

$$\tilde{F}_\pm = \frac{\mp R_0^3 \rho_2^3 a}{U^3(\rho_1 + \rho_2)} \int \frac{\omega^3(\omega \mp \omega_0)^2 \exp(-2\omega^2 h/g)}{\sqrt{\omega^2 - (g'^2/U^2)[1 \mp (\omega_0/\omega)^2]}}\, d\omega, \qquad (6.38)$$

where $g' = g \cdot \Delta\rho/(\rho_1 + \rho_2)$, R_0 is the radius of the ball, a is the displacement amplitude, and the region of integration is specified by the conditions: $\omega > 0$, $\omega^2 \geq (g'/U)^2(1 \mp \omega_0/\omega)^2$. The component $\tilde{F}_+$ is linked with the radiation of positive energy waves (in the reference system moving with the stretching velocity U) outside the "Cherenkov" cone, and the component $\tilde{F}_-$, with the NEW radiation inside the cone. The resultant effect depends on the stretching velocity (Froude's number $\mathrm{Fr} = U/\sqrt{gh}$) and on the oscillatory frequency (dimensionless parameter $s = 4\omega_0 U/g'$). Results of the numerical calculations of the oscillatory resistance force $\tilde{F}$ normalized to $f = (2\pi R^3)^2 \rho_2^2 a g'/2\pi h^4(\rho_1 + \rho_2)$ are presented in Fig. 6.11. The ball's oscillations become unstable in the region where $\tilde{F} < 0$, which gives rise to build-up of self-excited oscillations.

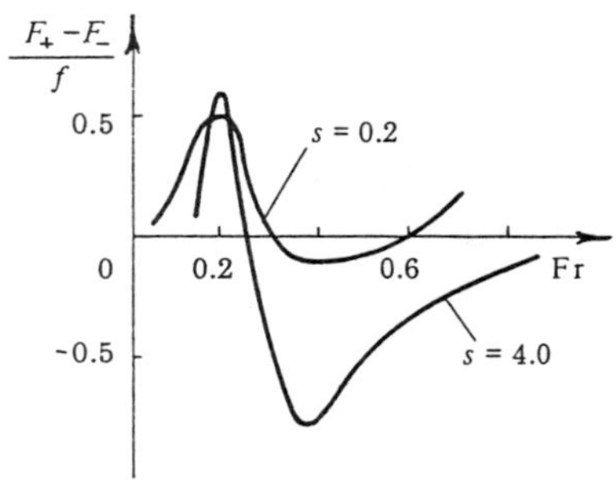

Figure 6.11: The amplitude of the normalized oscillatory force of resistance versus the Froude number Fr.

The limiting case of low-frequency oscillations ($s \ll 1$) when the radiated waves are concentrated near the "Cherenkov" cone deserves special attention. In this case, the force components $\tilde{F}_{\pm}$ have almost equal absolute values and the resulting force $\tilde{F}$ may be determined from the dependence of the constant component of force, F_0, on velocity U. Indeed, at sufficiently slow oscillations the radiative force is changing quasistatically, i. e., $F = F_0(U - d\xi/dt)$ and the oscillatory portion of the force is described by a variety of this expression

$$\tilde{F} = -\frac{\partial F_0}{\partial U}\frac{\partial \xi}{\partial t}. \tag{6.39}$$

When $\partial F_0/\partial U < 0$, the radiation into the region of anomalous Dopler effect (inside the "Cherenkov" cone) is more intense than in the region of normal Dopler effect ($F_- > F_+$), the work done by the force is greater than zero, and the oscillations become unstable.

The solution to the problem of the wave resistance of a smoothly moving body is well known in the literature (see, e. g., Sretensky, 1977). The dependence $F_0(U)$, actually, has a dropping portion at large velocities (Froude numbers) as a rule. At the same time, one can see from Fig. 6.11 that at high frequencies ($s \geq 1$) the instability may also occur in a broader region, outside the dropping portion of the curve.

The influence of the dissipation of radiated waves (the viscosity in the upper layer of the fluid is taken into account) on the radiative instability of oscillator as well as the nonlinear effects which determine the amplitude of self-excited oscillations were investigated by Dolina (1984). We would also like to mention the paper by Abramovich *et al.* (1986) where the radiative instablity of an oscillator moving in a continuously stratified fluid was analysed.

Chapter 7

Waves on Concentrated Vortices

Concentrated vortices, i. e., localized flows with closed streamlines, is a reference model broadly used in many hydrodynamical phenomena (Lamb, 1932). Such vortices give a good approximation for various real flows occurring both in nature (cyclonic and other atmospheric vortices, whirlpools, etc.) and as a result of the operation of different devices.

Of great interest is investigation of the oscillatory and wave disturbances on concentrated vortices. Vortex oscillations were considered in connection with the problem of circular wave shedding behind an airplane wing (Hall, 1972; Uberoi, 1979), tornados (Rotunno, 1978; Snow, 1978), in investigations of the instability of rotating jets (Uberoi *et al.*, 1972; Lessen *et al.*, 1973) and of the properties of a superfluid helium (Fetter and Harvey, 1971; Sonin, 1975; Ignat'ev and Sonin, 1981), as well as for the interpretation of the peculiarities of neutron stars rotation (Tsakadze J. and Tsakadze S., 1975; Pains, 1980; Alpar *et al.*, 1981).

Localized vortices often serve as an energy source for different wave motions. In particular, they underlie some mechanisms of aerodynamic generation of sound (Klyatskin, 1966; Howe, 1976; Mohring, 1978; Gryanik, 1983; Kambe *et al.*, 1985; Ribner, 1985; Lyamshev and

Skvortsov, 1988). The interactions of vortices with sound, internal gravity waves, inertial and other types of waves in a stratified medium may prove to be a significant element of the dynamics of many phenomena observed in nature (Willoughby, 1977; Gostintsev *et al.*, 1986; Shchepetnov *et al.*, 1986; Gavrilov, 1987; Korotaev, 1988).

The theory of small oscillations of cylindrical vortices has much in common with the corresponding theory for plane-parallel flows. Here, by virtue of a cylindrical symmetry, it is convenient to resolve the perturbation field into cylindrical harmonics. Some methods used for the investigation of cylindrical waves on axially symmetric vortices and possible mechanisms of the main flow–wave interaction analogous to the ones for plane-parallel flows will be considered in this chapter.

7.1 An algebraic method for columnar vortices

Investigations of small perturbations in axially symmetric flows, similarly to plane-parallel shear flows, are often performed with the use of an algebraic method based on the approximation of the velocity distribution by a profile with piecewise vorticity and on matching the analytical solutions at the boundaries of the regions. Different variants of this method were employed for the analysis of the stability of cylindrical vortices and jets [among the pioneers in this field we can name Ponstein (1959), Howard and Gupta (1962), Michalke and Timme (1967), Busse (1968)]. In this chapter a generalization of the algebraic method and its possible applications in different wave problems are considered.

The equations for small perturbations of a two-dimensional flow in a plane-laminar rotating medium have the following form

$$\begin{aligned} \frac{\partial \mathbf{v}}{\partial t} + (\mathbf{U}\nabla)\mathbf{v} + (\mathbf{v}\nabla)\mathbf{U} + 2[\mathbf{\Omega}_0, \mathbf{v}] + \frac{\nabla p}{\rho_0} &= \nu\Delta\mathbf{v}, \\ \frac{\partial w}{\partial t} + (\mathbf{U}\nabla)w + \frac{1}{\rho_0}\frac{\partial p}{\partial z} + \frac{g\rho}{\rho_0} &= \nu\Delta w, \end{aligned} \qquad (7.1a)$$

$$\begin{aligned} \frac{\partial \rho}{\partial t} + (\mathbf{U}\nabla)\rho + w\frac{d\rho_0}{dz} + \rho_0\left[(\nabla\mathbf{v}) + \frac{\partial w}{\partial z}\right] &= 0, \\ \frac{\partial \rho}{\partial t} + (\mathbf{U}\nabla)\rho + w\frac{d\rho_0}{dz} - \frac{1}{c_s^2}\left[\frac{\partial p}{\partial t} + (\mathbf{U}\nabla)p - \rho_0 g w\right] &= 0. \end{aligned} \tag{7.1b}$$

Here, $\rho_0(z)$ is the unperturbed velocity profile, $\mathbf{\Omega}_0$ is the angular velocity of the medium rotation around the vertical axis, ρ and p are the density and pressure perturbations related by the local velocity as $p = c_s^2\rho$, w, and $\mathbf{v}$ are the vertical and horizontal components of small velocity perturbations with respect to the unperturbed horizontal flow whose velocity $\mathbf{U} = (U_x(x,y),\ U_y(x,y),\ 0)$ is independent of the vertical coordinate z. The model considered here allows for the investigation of waves in the barotropic flows of a stratified medium where the density is inhomogeneous along the vertical and the velocity vector is in the horizontal plane. By passing to the field variables

$$\tilde{\mathbf{v}} = \mathbf{v}\sqrt{\frac{\rho_0(z)}{\rho_0(0)}}, \quad \tilde{w} = w\sqrt{\frac{\rho_0(z)}{\rho_0(0)}}, \quad \tilde{\rho} = \rho\sqrt{\frac{\rho_0(0)}{\rho_0(z)}}, \quad \tilde{p} = p\sqrt{\frac{\rho_0(0)}{\rho_0(z)}}, \tag{7.2}$$

one obtains from (7.1) for the amplitudes of oscillations

$$\left\{\begin{array}{c} \tilde{\mathbf{v}} \\ \tilde{w} \\ \tilde{\rho} \\ \tilde{p} \end{array}\right\} = \left\{\begin{array}{c} \mathbf{v} \\ w \\ b\rho_0(0) \\ p \end{array}\right\} e^{-i\omega t}$$

the following system of equations

$$\begin{aligned} i\omega\mathbf{v} + (\mathbf{U}\nabla)\mathbf{v} + (\mathbf{v}\nabla)\mathbf{U} + 2\left[\mathbf{\Omega}_0, \mathbf{v}\right] + \nabla p &= \nu\Delta\mathbf{v}, \\ i\omega w + (\mathbf{U}\nabla)w + \frac{\partial p}{\partial z} + gb + \frac{d\rho_0}{dz}\frac{p}{2\rho_0} &= \nu\Delta w, \\ i\omega b + (\mathbf{U}\nabla)b + \frac{w}{\rho_0}\frac{d\rho_0}{dz} + (\nabla\mathbf{v}) + \frac{\partial w}{\partial z} - \frac{d\rho_0}{dz}\frac{w}{2\rho_0} &= 0, \\ i\omega b + (\mathbf{U}\nabla)b + \frac{d\rho_0}{dz}\frac{w}{\rho_0} - \frac{-i\omega p + (\mathbf{U}\nabla)p - gw}{c_s^2} &= 0. \end{aligned} \tag{7.3}$$

In Chap. 8 we will consider the IGW scattering at vortices of arbitrary configuration within the frames of the Born approximation when the unperturbed flow has a rather small velocity (low-intensity vortices). This approximation does not hold for the investigation of the intrinsic oscillations of vortices, neither is it valid for the resonant phenomena when the flow velocity is comparable with the corresponding phase velocity of the wave. No analytical solutions can be constructed for vortex flows in a general form but one can do it in the case of the wave scattering in axially symmetric flows with velocity $\mathbf{V} = [\mathbf{r}, \mathbf{\Omega}]$, the angular velocity $\mathbf{\Omega}(r)$ being independent of the azimuthal coordinate φ. In the cylindrical coordinates, (7.3) yields the equations for the amplitudes of harmonics of the small disturbances proportional to $e^{in\varphi+iqz}$:

$$
\begin{aligned}
-i\kappa u + 2(\Omega+\Omega_0)v + rv\frac{d\Omega}{dr} + \frac{inp}{r} &= \nu\left(\tilde{H}u - \frac{u}{r^2} - uq^2 + \frac{2in}{r}v\right), \\
-i\kappa v - 2(\Omega+\Omega_0)u + \frac{dp}{dr} &= \nu\left(\tilde{H}v - \frac{v}{r^2} - vq^2 - \frac{2in}{r}u\right), \\
-i\kappa\left(1 - \frac{N^2}{\kappa^2}\right)w + (iq-\Gamma)p &= \nu(\tilde{H}w - wq^2), \qquad (7.4) \\
-\frac{i\kappa}{c_s^2}p + \frac{dv}{dr} + \frac{v}{r} + \frac{in}{r}u + (iq+\Gamma)w &= 0, \\
b &= \frac{p}{c_s^2} + \frac{iN^2 w}{\kappa g}.
\end{aligned}
$$

Here, operator $\tilde{H} = \frac{d^2}{dr^2} + \frac{1}{r}\frac{d}{dr} - \frac{n^2}{r^2}$, $\kappa = \omega - n\Omega(r)$, u and v are the amplitudes of the azimuthal and radial velocity components, and the parameters N and Γ are taken to be constant. Neglecting viscosity ($\nu = 0$), one can substitute instead of the vertical velocity component

$$
w = -\frac{ip}{\kappa}\frac{\Gamma + iq}{1 - N^2/\kappa^2}
$$

and find from (7.4) the system of equations

$$
\begin{aligned}
-i\kappa u + Qv + \frac{inp}{r} &= 0, \\
-i\kappa v - 2(\Omega + \Omega_0)v + \frac{dp}{dr} &= 0, \qquad (7.5) \\
\frac{i\kappa p}{\tilde{c}^2} + \frac{dv}{dr} + \frac{v}{r} + \frac{inu}{r} &= 0,
\end{aligned}
$$

where $Q = rd\Omega/dr + 2(\Omega + \Omega_0)$ is absolute vorticity. The system (7.5) coincides with the system of equations for the two-dimensional perturbations of a cylindrical vortex in a compressible medium without stratification (Golemshtok and Fabrikant, 1980). Here, however, the quantity

$$\tilde{c}^{-2} = \frac{1}{c_s^2} + \frac{q^2 + \Gamma^2}{N^2 - \kappa^2}$$

is not constant, generally speaking, and may depend on the wave frequency and flow velocity.

Finally, by expressing the velocity perturbations through pressure perturbations

$$u = \frac{1}{F}\left(\frac{\kappa np}{r} - Q\frac{dp}{dr}\right), \qquad v = \left(2in(\Omega + \Omega_0)\frac{p}{r} - i\kappa\frac{dp}{dr}\right), \qquad (7.6)$$

where $F = \kappa^2 - 2Q(\Omega + \Omega_0)$, one can transform (7.5) to one equation

$$\tilde{H}p - \frac{F'(r)}{F(r)}\frac{dp}{dr} + pF\left(\frac{1}{\tilde{c}^2} - \frac{2n}{\kappa r}\frac{d}{dr}\frac{\Omega + \Omega_0}{F}\right) = 0. \qquad (7.7)$$

The parameter $\tilde{c}$ in (7.5) and (7.7) has the meaning of the horizontal phase velocity of the wave in a motionless medium in the absence of rotation: if $\Omega = \Omega_0 = 0$, then

$$\tilde{c}^{-2} = \frac{1}{c^2} + \frac{q^2 + \Gamma^2}{N^2 - \omega^2}. \qquad (7.8)$$

For $\omega \gg N$ and $\Omega \gg N$, the system of equations (7.5) describes the propagation of perturbations in a compressible homogeneous medium.

The vortex oscillations in this limit were investigated by Golemshtok and Fabrikant (1980). In another relatively simple case

$$\Omega_0 \ll \omega \ll N, \quad \Omega_0 \ll \Omega \ll N, \tag{7.9}$$

one has $\tilde{c} = \text{const}$ and the problem reduces to the particular case of two-dimensional perturbations which were also analysed by Golemshtok and Fabrikant (1980).

Equation (7.7) is simplified significantly in an homogeneous incompressible medium for $\Omega_0 = 0$. In two particular cases it transforms to the modified Bessel's equations:

for a potential flow ($Q = 0$, $\Omega \sim r^{-2}$):

$$\tilde{H}(p/\kappa) - q^2(p/\kappa) = 0, \tag{7.10}$$

and for an homogeneous vortex flow ($\Omega = \text{const}$):

$$\tilde{H}p - q^2(1 - 4\Omega^2/\kappa^2)p = 0. \tag{7.11}$$

For two-dimensional perturbations ($q = 0$), (7.7) in this approximation gives an equation for the stream function ψ

$$\tilde{H}\psi + \frac{nQ'(r)}{\kappa r}\psi = 0 \tag{7.12}$$

through which the velocity and pressure amplitudes are expressed:

$$u = \frac{d\psi}{dr}, \quad v = \frac{in\psi}{r}, \quad p = \rho\left(\frac{\kappa r}{n}\frac{d\psi}{dr} + Q\psi\right).$$

The general solution to (7.12) can be found only for the flows with constant or zero vorticity for which $Q'(r) = 0$ and the angular velocity $\Omega(r) = B + A/r$. If $A = 0$ or $B = 0$, then not only (7.12) but also a more general, three-dimensional problem may be solved by employing (7.10) or (7.11). A continuous profile of angular velocity is approximated by the model profile that consists of several portions of uniform vorticity ($Q'(r) = 0$) separated by vortex sheets or velocity kinks. Small perturbations for such a "piecewise" profile may be investigated by "matching" the solutions in each portion using the boundary conditions at the discontinuities. By analogy with plane-parallel flows, when the Rayleigh equation is solved analytically for piecewise-linear velocity profiles, such a "piecewise" approximation with subsequent matching of exact solutions at discontinuities may be called an algebraic method (Betchov and Criminale, 1967).

The boundary conditions on a vortex sheet at $r = a + \zeta$ (where ζ is the displacement of the oscillating boundary relative to the equilibrium position $r = a$) is the continuity at the moving pressure boundary and at the velocity component normal to this boundary. These conditions for the amplitudes of perturbations are written as

$$\left[p + \zeta \frac{\partial P}{\partial r}\right]_{r=a-0}^{r=a+0} = \left[p + \rho\Omega^2\zeta\right]_{r=a-0}^{r=a+0}, \qquad \left[\frac{v}{\kappa}\right]_{r=a-0}^{r=a+0} = 0. \tag{7.13}$$

Using these boundary conditions one can match the solutions to (7.10) or (7.11) in the neighboring flow regions for the amplitudes of the sought functions. These amplitudes for a potential flow have the following form:

$$\begin{aligned} u &= D\frac{nW_n(qr)}{r}, \qquad & v &= iD\frac{\partial W_n(qr)}{\partial r}, \\ w &= DqW_n(qr), \qquad & p &= D\kappa\rho_0 W_n(qr). \end{aligned} \tag{7.14}$$

The corresponding equations for a flow of uniform vorticity with $\Omega = \text{const}$ are written as

$$\begin{aligned} u &= \frac{1}{s^2\kappa^2} D\left[\kappa n \frac{W_n(sqr)}{r} - 2\Omega \frac{\partial W_n(sqr)}{\partial r}\right], \\ v &= \frac{i}{s^2\kappa^2} D\left[\kappa \frac{\partial W_n(sqr)}{\partial r} - 2n\Omega \frac{W_n(sqr)}{r}\right], \\ w &= \frac{q}{\kappa} DW_n(sqr), \\ p &= D\rho_0 W_n(qr), \end{aligned} \tag{7.15}$$

where $s = \sqrt{1 - 4\Omega^2/\kappa^2}$ and W_n are the modified Bessel functions.

Consider the simplest "piecewise" velocity profiles which contain only one discontinuity separating the outer region of the potential flow (where $Q = 0$) and the vortex core with a potential (Figs. 7.1a,b) or uniformly rotating ($\Omega = \text{const}$, Figs. 7.1c,d) flow.

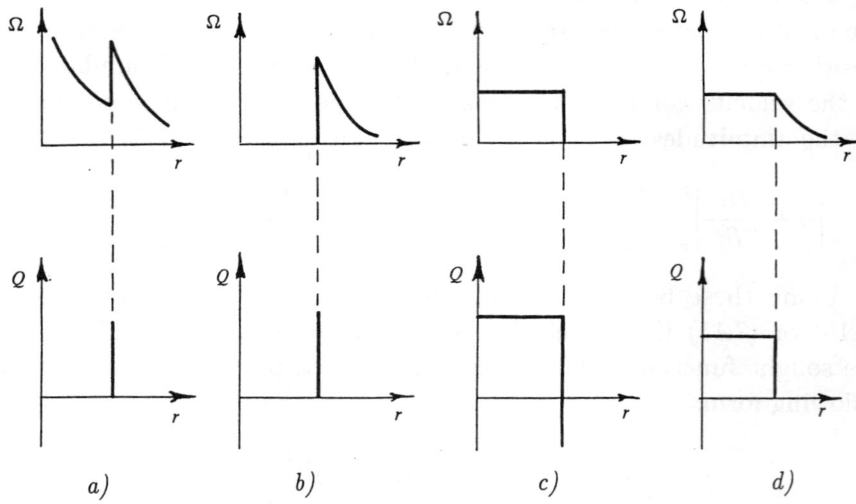

Figure 7.1: The profiles of angular velocity $\Omega(r)$ and vorticity $Q(r)$ for cylindrical vortices: (a) vortex with a potential core, (b) cylindrical vortex sheet, (c) rotating liquid cylinder, (d) Rankine vortex.

Investigation of the stability of such flows in a perfect fluid is an analog of the Kelvin–Helmholtz problem for circular flows. Note that the results that will be presented below can easily be generalized to the flows with the vertical velocity $V_z = U = \text{const}$. To this end, one only needs to take into account the Dopler shift in the expression for $\kappa = \omega - n\Omega - qU$ in the dispersion equation and in other expressions [see (Uberoi *et al.*, 1972; Lessen *et al.*, 1973) where the effect of rotation on the stability of liquid jets was analysed]. The capillary forces on the discontinuity (Uberoi *et al.*, 1972) are also readily taken into account.

The solutions to (7.10) and (7.11) must be chosen with allowances made for the boundedness of intrinsic solutions at $r \to 0$, $r \to \infty$. Then, the solution in the inner region (core) of the vortex is expressed by the modified Bessel function of the first kind I_n, and in the outer region by the functions of the second kind K_n. By using an algebraic method one can readily obtain a dispersion equation for the eigenvalues of the frequency ω of perturbations in piecewise-potential flows (Figs. 7.1a,b):

$$\rho_i \left[\kappa_i^2 Z_{in}(qa) + \Omega_i^2\right] = \rho_e \left[\kappa_e^2 Z_{en}(qa) + \Omega_e^2\right], \tag{7.16}$$

where the subscripts i and e designate the corresponding values in the inner and outer regions, in particular,

$$Z_{in}(qa) = \left.\frac{\mathrm{I}_n(qr)}{r d\mathrm{I}_n(qr)/dr}\right|_{r=a}, \qquad Z_{en}(qa) = \left.\frac{\mathrm{K}_n(qr)}{r d\mathrm{K}_n(qr)/dr}\right|_{r=a},$$

$$\Omega_i = \Omega|_{r=a-0}, \quad \Omega_e = \Omega|_{r=a+0}.$$

Some partial solutions to (7.16) were investigated by Uberoi *et al.* (1972), Lessen *et al.* (1973), and Rotunno (1978) [see also (Michalke and Timme (1967), Zakharov (1976); these authors, however, used incorrect boundary conditions at the discontinuity]. Consider now two-dimensional ($q = 0$) oscillations of different vortices and axially symmetric ($n = 0$) waves travelling along the vortex axis.

The solution of (7.16) for piecewise-potential flows (Figs. 7.1a,b) is written in the form

$$\omega = \frac{n(\rho_i \Omega_i Z_{in} - \rho_e \Omega_e Z_{en}) \pm Sq}{\rho_i Z_{in} - \rho_e Z_{en}} \tag{7.17}$$

where

$$Sq = \sqrt{n \rho_e \rho_i Z_{in} Z_{en} (\Omega_e - \Omega_i)^2 - (\rho_i Z_{in} - \rho_e Z_{en})(\rho_i \Omega_i^2 - \rho_e \Omega_e^2)}.$$

Assuming $\rho_i = \rho_e$, for the axially symmetric mode with $n = 0$ one finds

$$\omega = qa\sqrt{(\Omega_i^2 - \Omega_e^2)\mathrm{I}_1(qa)\mathrm{K}_1(qa)} \underset{qa \to 0}{\longrightarrow} qa\sqrt{\frac{1}{2}(\Omega_i^2 - \Omega_e^2)}. \tag{7.18}$$

One can see that this mode is unstable if $\Omega_i > \Omega_e$. Otherwise ($\Omega_i < \Omega_e$), it is a wave of symmetric pulsations (Fig. 7.2) travelling along the axis of the vortex with high-frequency dispersion (Fig. 7.3).

Consider also the oscillations of the cylindrical density discontinuity. If the angular velocity is continuous ($\Omega_i = \Omega_e = \Omega$), the frequency is determined by

$$\omega = \Omega n \pm \Omega\sqrt{\frac{\rho_e - \rho_i}{\rho_i Z_{in} - \rho_e Z_{en}}} \underset{qa\to 0}{\longrightarrow} \Omega n \pm \Omega\sqrt{n\frac{\rho_e - \rho_i}{\rho_e + \rho_i}}. \tag{7.19}$$

Such oscillations are similar to the oscillations of the surface of heavy liquid but the centrifugal force, instead of the gravity force, is the return force in this case. The instability arising at $\rho_i > \rho_e$ is analogous to the Rayleigh–Taylor instability.

Finally, one can write for the two-dimensional ($q = 0$) oscillations with $n \geq 1$ (Fig. 7.2) the following dispersion equation

$$\rho_i\left[\kappa_i(\kappa_i + 2\Omega_i) + |n|\Omega_i^2\right] + \rho_e(\kappa_e^2 - |n|\Omega_e^2) = 0, \tag{7.20}$$

the solutions to which at $\rho_i = \rho_e$ have the form

$$\omega = \frac{\Omega_e + \Omega_i}{2}\left[n \pm \sqrt{|n|\Delta(2 - |n|\Delta)}\right], \tag{7.21}$$

where $\Delta = (\Omega_e - \Omega_i)/(\Omega_e + \Omega_i)$. Such oscillations are unstable at any value of n, if $\Delta < 0$ or $\Delta > 2$. Whereas for $0 < \Delta < 2$, only the harmonics with $|n| \leq 2/\Delta$ are stable. Of particular interest is the flow with $\Omega_i = 0$, i. e., a cylindrical vortex sheet (Fig. 7.1b). Its stability (an exact cylindrical analog of the Kelvin–Helmholtz problem) is determined by the frequencies of the harmonic $n \geq 1$ at $\Delta = 1$:

$$\omega = \frac{\Omega_e}{2}\left(n \pm \sqrt{2|n| - n^2}\right). \tag{7.22}$$

Here, the harmonics with $n \geq 3$ are unstable and, as n is increasing, the solution to (7.22) tends to the solution for a plane vortex sheet:

$$\omega = (1 + i)n\Omega_e/2.$$

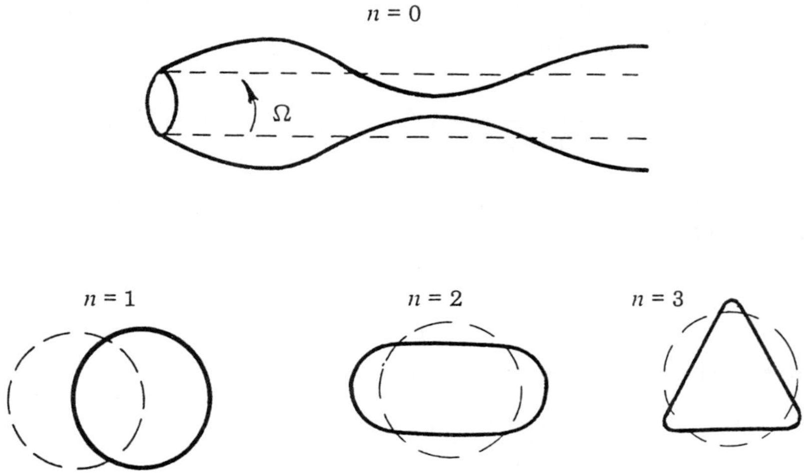

Figure 7.2: Cylindrical vortex modes.

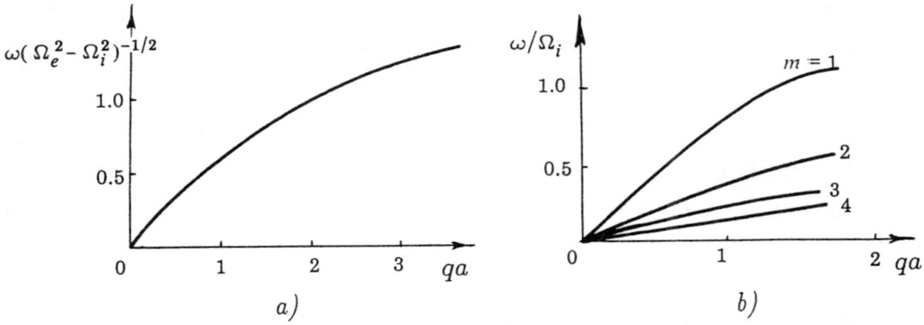

Figure 7.3: Dispersion dependences for the axially-symmetric waves travelling along the axis of a potential vortex (a) and along the axis of Rankine vortex (b).

The dispersion equation for vortices with a solid-state rotating core (Fig. 7.1c,d) has the form

$$\rho_i\left[\frac{\kappa_i^2+\Omega_i^2}{Z_{in}^{-1}(qas)-2n\Omega_i/\kappa_i}+\Omega_i^2\right]=\rho_e[\kappa_e^2 Z_{en}(qa)+\Omega_e^2]. \tag{7.23}$$

For the axially symmetric waves along the Rankine vortex with the uniform density $\rho_e=\rho_i$ and uniform rotation $\Omega_e=\Omega_i=\Omega$ (Fig. 7.1d) one has

$$\omega=\frac{2qa\Omega}{\sqrt{p^2+q^2a^2}}, \tag{7.24}$$

where p is the root of the transcendental equation

$$\frac{p\mathrm{J}_0(p)}{\mathrm{J}_1(p)}+\frac{qa\mathrm{K}_0(qa)}{\mathrm{K}_1(qa)}=0. \tag{7.25}$$

Here, $\mathrm{J}_n(z)$ is the Bessel function of the first kind. The long waves ($qa\to 0$) have a countable number of dispersion branches (Fig. 7.3):

$$\omega=2\Omega_i qa/p_m, \tag{7.26}$$

where p_m are the zeros of the function $\mathrm{J}_0(p)$.

At $\rho_e\neq\rho_i$, $\Omega_e\neq\Omega_i$, the dispersion equation for long axially symmetric waves also has the form (7.26) but p_m are the roots of a different equation

$$\frac{\mathrm{J}_0(p)}{\mathrm{J}_1(p)}=\frac{p}{4}\left(\frac{\rho_e\Omega_e^2}{\rho_i\Omega_i^2}-1\right).$$

The family of self-excited oscillations with different m are, actually, inertial waves, i. e., the eigenmodes of the waveguide formed by the cylindrical vortex sheet.

The two-dimensional ($k=0$) vortex oscillations with a uniformly rotating core, the dispersion equation for which is written as

$$\rho_i[\kappa_i(\kappa_i+2\Omega_i)+|n|\Omega_i^2]+\rho_e[\kappa_e^2-|n|\Omega_e^2]=0, \tag{7.27}$$

have at $\rho_e=\rho_i$ the following frequencies

$$\omega=\frac{1}{2}\left[(n-1)\Omega_i+n\Omega_e\pm Sr\right], \tag{7.28}$$

where $Sr = \sqrt{n(1-n)(\Omega_e - \Omega_i)^2 + \Omega_i^2(n - 2|n| + 1) + \Omega_e^2(2|n| - n)}$.

For the Rankine vortex ($\Omega_e = \Omega_i = \Omega$) whose stability was shown already by Kelvin (Lamb, 1932), the frequencies of two-dimensional intrinsic oscillations are equal to

$$\omega_1 = \Omega(n-1), \quad \omega_2 = \Omega n. \tag{7.29}$$

Consider the effect of viscosity inside the vortex core on the stability of wave perturbations. This can be done in the frames of the algebraic method restricting the investigation to "piecewise" velocity profiles. For this purpose consider, like it was done in Sect. 4.3, the discontinuity between the viscous and inviscid media, i. e., take into account the viscosity only in the core of a cylindrical vortex.

The solutions to (7.10) for small perturbations in a potential flow of a perfect fluid satisfy identically a more general system of equations (7.4) in a quasistatic approximation when $\tilde{c} \to \infty$. This approximation is valid, in particular, for an incompressible homogeneous viscous fluid, provided $Q = 0$. For an uniformly rotating liquid ($\Omega = \text{const}$), two-dimensional inviscid oscillations, i. e., the solutions of (7.4) at $k \to 0$, satisfy identically the system (7.4) where $\tilde{c} = \infty$. Thus, allowance for viscosity does not change the solutions obtained for small oscillations in a perfect fluid in these two particular cases. Whereas an increase of the order of equations, with viscous dissipation taken into account, leads to the appearance of additional linearly independent solutions which rapidly oscillate and damp at large Reynolds numbers. These solutions will be investigated assuming in (7.4) $c_s = \infty$ and $N = \infty$ and retaining only the higher-order terms with respect to the Reynolds number:

$$\begin{aligned} i\kappa u &= \nu \frac{d^2 u}{dr^2}, & -i\kappa w &= \nu \frac{d^2 w}{dr^2}, \\ \frac{dv}{dr} &= -\frac{inu}{r} - ikw, & \frac{dp}{dr} &= 2\rho\Omega u. \end{aligned} \tag{7.30}$$

The solution rapidly decaying with increasing distance from the boundary $r = a$ is written as

$$\begin{aligned} u &= D\tilde{\varepsilon}, \qquad v = \left(\frac{in}{k_\nu a}D + \frac{iq}{k_\nu}G\right)\tilde{\varepsilon}, \\ w &= G\tilde{\varepsilon}, \qquad p = \frac{2\rho\Omega}{k_\nu}D\tilde{\varepsilon}, \end{aligned} \tag{7.31}$$

where $\tilde{\varepsilon} = \exp[k_\nu(r-a)]$, $k_\nu = \sqrt{-i\kappa/\nu 2}$ (with $\mathrm{Re}\, k_\nu > 0$), while D and G are arbitrary constants.

The dynamical boundary conditions are the continuity of the flow pulse components normal to the moving boundary, i. e., of the components $\sigma_{ik}n_k$ (where n is a unit vector normal to the discontinuity) of the stress tensor (Landau and Lifshits, 1986). In a linear approximation with respect to the oscillatory amplitude, the following quantities may be considered to be continuous at the unperturbed boundary $r = a$:

$$\begin{aligned} \sigma_{rr} + \zeta\frac{\partial\sigma_{rr}^{(0)}}{\partial r} - \frac{\sigma_{r\varphi}^{(0)}}{a}\frac{\partial\zeta}{\partial\varphi} &= \left(-p + 2\rho\nu\frac{dv}{dr} - \rho a\zeta\Omega^2 - in\rho\nu\zeta\frac{d\Omega}{dr}\right)\tilde{E}, \\ \sigma_{rz} + \zeta\frac{\partial\sigma_{rz}^{(0)}}{\partial r} - \frac{\sigma_{z\varphi}^{(0)}}{a}\frac{\partial\zeta}{\partial\varphi} &= \left(\frac{dw}{dr} + ikw\right)\rho\nu\tilde{E}, \\ \sigma_{r\varphi} + \zeta\frac{\partial\sigma_{r\varphi}^{(0)}}{\partial r} - \frac{\sigma_{\varphi\varphi}^{(0)}}{a}\frac{\partial\zeta}{\partial\varphi} &= \left[\frac{du}{dr} - \frac{u}{r} + \frac{inv}{r} + \zeta\left(r\frac{d^2\Omega}{dr^2} + \frac{d\Omega}{dr}\right)\right]\rho\nu\tilde{E}. \end{aligned} \tag{7.32}$$

Here, $\tilde{E} = \exp(-i\omega t + in\varphi + ikz)$ and $\sigma_{ik}^{(0)}$ is the unperturbed stress tensor.

These boundary conditions together with the kinematic condition (the continuity of the quantity v/κ) allow for matching the inviscid solutions of (7.10) in the outer region of the vortex and the linear combination of inviscid and damping (7.31) solutions for the oscillations in the core. One can readily obtain for piecewise-potential flows, instead of (7.16), a dispersion equation taking into account small viscosity

$$\rho_i\left[\kappa_i^2 Z_{in}(qa) + \Omega_i^2\right] - \rho_e\left[\kappa_e^2 Z_{en}(qa) + \Omega_e^2\right]$$

$$= \frac{2i\rho_i\nu}{\kappa a^2}\left(\kappa_i^2 + 4\Omega_i^2 + 3n\kappa_i\Omega_i + n^2\kappa_i^2 Z_{in}^2\right. \tag{7.33}$$
$$\left. - 2n^2\kappa_i^2 Z_{in} - 4in\Omega_i\kappa_i Z_{in} - 2q^2a^2\kappa_i^2 Z_{in}\right).$$

The dispersion equation for the two-dimensional perturbations on vortices with a solid-state rotating core in the presence of small viscosity inside can be derived analogously

$$\rho_e\left(\kappa_e^2 - n\Omega_e^2\right) + \rho_i\left[\kappa_i(\kappa_i + 2\Omega_i) + n\Omega_i^2\right] = \frac{4i\nu}{a^2}\rho_i n(n-1)\kappa_i. \tag{7.34}$$

For the Rankine vortex (Fig. 7.1d), the frequencies of the intrinsic two-dimensional oscillations in the first approximation with respect to the parameter $\mathrm{Re} = \nu/\Omega a^2$ have the form

$$\omega = (n-1)\Omega - 2i\nu n(n-1)/a^2. \tag{7.35}$$

These frequencies correspond to the damping oscillations. In the interpretation of this result one must take into account that the dispersion is negative in the medium rotating faster than the angular phase velocity of azimuthal waves $(\omega - n\Omega < 0)$. However, the energy of vortex oscillations is also negative in this case (Kop'ev and Leont'ev, 1983):

$$E = -\pi a^4\Omega^2\frac{n-1}{n}\rho\left|\frac{\zeta}{a}\right|^2 < 0, \tag{7.36}$$

where ζ is the amplitude of displacement of the vortex core boundaries, which results in damping oscillations.

The negative energy of the intrinsic oscillations of the vortex makes its radiative instability possible in a compressible medium (Broadbent and Moore, 1979). For $M = \Omega a/\tilde{c} \ll 1$, the oscillations are close to quasistatic (incompressible) ones at $r \leq a$ but the nonstationary flow associated with it radiates waves in the wave zone $r \geq \lambda$. The amplitude of oscillations is growing as the energy is taken away by the departing waves. Note that the instability caused by the radiative losses is known in other physical systems too: for the beams of charged particles in a plasma (Erokhin and Moiseev, 1973), in inverted media (Zheleznyakov *et al.*, 1984), etc.

It is worthy of notice that the radiative acoustic vortex instability is asssociated with the presence in an incompressible approximation of a neutrally stable mode. This mode exists in the Rankine vortex at the "kink" of the vorticity profile. Whereas for the smooth vorticity profile, the resonant absorption in a critical layer leads to a damping eigenmode of the vortex (Danilov, 1989). In this case, weak compressibility cannot give rise to the radiative instability of the vortex.

7.2 Wave amplification by a vortex with a viscous core: Zel'dovich's effect in hydrodynamics

The presence of negative energy waves in a rotating medium allows, in principle, for the amplification of the waves incident on the vortex. However, the necessary condition here is a sink for NEWs. Such a sink in plane-parallel vortex sheets is vanishing of the refracted waves to the infinity. It is impossible for the vortices having a finite core region. However, the scattered wave may be amplified due to the changes of the sign of viscous dissipation in the vortex core.

Consider the interaction of a wave with a cylindrical vortex in a quasistatic approximation: $\tilde{\Omega} \leq \omega \ll \tilde{c}/a$, when the vortex with the characteristic frequency $\tilde{\Omega}$ and characteristic size of the vortex core a lies completely in the non-wave zone.

Let us calculate the value of the Reynolds stress that describes the flow of the tangential pulse component through the cylindrical surface surrounding the vortex: $L(r) = \mathrm{Re}\,(r^2 u^* v)$. The flow is potential ($u = in\Phi/r$, $v = d\Phi/dr$) outside the vortex core at $r \gg a$ and (7.5) yields, at $\Omega_0 = 0$, for the potential Φ

$$\frac{d^2\Phi}{dr^2} + \frac{1}{r}\frac{d\Phi}{dr} + \left(\frac{\kappa^2}{\tilde{c}^2} - \frac{n^2}{r^2}\right)\Phi = 0. \tag{7.37}$$

Multiplication of equality (7.37) by $inr\Phi^*$ and integration by parts from $r = r_1$ to $r = r_0$ readily show that the equality $L(r_0) = L(r_1)$

take place, where $a \ll r_1 \ll \lambda$ and $r_0 \gg \lambda = 2\pi\tilde{c}/\omega$. The approximate solution to (7.37), at $\mu = \omega a/\tilde{c} \ll 1$, is expressed by the Hankel functions of the first and second kinds

$$\Phi = A\left[\mathrm{H}_n^{(1)}(\omega r/\tilde{c}) + R_n \mathrm{H}_n^{(2)}(\omega r/\tilde{c})\right], \tag{7.38}$$

where R_n is the complex reflection coefficient of the cylindrical harmonic. From this follows

$$L(r_0) = \frac{2|n||A|^2}{\pi}(|R_n|^2 - 1), \tag{7.39}$$

i. e., the direction of the pulse flow in the wave zone depends on how much the absolute value of the reflection coefficient differs from zero.

At the same time, (7.12) for the stream function amplitude may be used in the quasistatic region at $r \ll |n|\tilde{c}/\omega$. The solution to (7.12) in the intermediate region $a/p \ll \lambda$ takes the form

$$\psi = B\left[(r/a)^{|n|} + q_n (r/a)^{-|n|}\right], \quad B = \mathrm{const.} \tag{7.40}$$

Comparison of this solution with the asymptotic behavior of the solution (7.38) in the region $r \ll \lambda$ shows that

$$1 - R_n \leq \mu^2 \quad \text{and} \quad |B| = \frac{2}{|n|!}\left(\frac{\mu}{2}\right)^n |A|.$$

Besides, from (7.40) follows $L(r_1) = -2n^2|E|^2 \mathrm{Im}\, q_n$. This, with (7.39) taken into account, yields

$$1 - |R_n|^2 = \frac{4\pi \mathrm{Im}\, q_n}{|n|!(|n|-1)!}\left(\frac{\omega a}{2\tilde{c}}\right)^{2|n|}. \tag{7.41}$$

The values of q_n for the flows with piecewise-constant vorticity can be calculated by means of the algebraic method considered above. Taking into account the viscosity inside the vortex core and matching the solutions at a cylindrical vortex sheet one obtains

$$1 - |R_n|^2 = \frac{16\pi\nu\kappa_i\kappa_e^2 S_n (\omega a/2\tilde{c})^{2|n|}}{|n|!(|n|-1)!a^2\left[\kappa_i T_n + |n|\Omega_i^2 + (\kappa_e^2 - |n|\Omega_e^2)\rho_e/\rho_i\right]^2}, \tag{7.42}$$

where $T_n = \kappa_i$ and $S_n = 2(n^2 - |n|) - (n\Omega_i/\kappa_i)(|n| - 4) - 4\,(n\Omega_i/\kappa_i)^2$ for the piecewise-potential flows (Fig. 7.1a,b) while $T_n = \kappa_i + 2\Omega_i n/|n|$ and $S_n = 2(n^2 - |n|)$ for the vortices with a uniformly rotating core (Fig. 7.1c,d).

The expression (7.41) describes the viscous absorption of a cylindrical wave by a vortex. For the vortices with an uniformly rotating core, its sign depends on the sign of the quantity $\kappa_i = \omega - n\Omega_i$: if $\kappa_i < 0$, the wave is amplified ($|R| > 1$). In particular, for the Rankine vortex one has

$$1 - |R_n|^2 = \frac{8\pi\nu}{(|n|-1)!(|n|-2)!a^2}\frac{\kappa}{(\kappa+\Omega)^2}\left(\frac{\omega a}{2\tilde{c}}\right)^{2|n|}. \qquad (7.43)$$

Wave amplification by a rotating liquid cylinder is the hydrodynamic analog of the effect considered by Ya. B. Zel'dovich (1972) who showed that electromagnetic waves may be amplified at scattering by a rotating conducting cylinder and gravity waves may be amplified on a collapsing rotating star (Starobinsky, 1973).

Note, however, that the mechanism of viscous dissipation considered above is quite specific and does not reduce merely to wave absorption. Such an absorption, for instance, does not occur for two-dimensional potential oscillations in a liquid with uniform vorticity. The dissipation of the waves refracted at the discontinuity is governed by the linear transformation to rapidly damping vortex oscillations.

7.3 Resonant wave amplification by vortices in a perfect fluid

The simple examples considered above are model ones and the found viscous solutions do not hold for continuous velocity profiles if the width of a shear layer, l, is not small compared to the viscous scale $l_\nu = (\nu/\kappa_i)^{1/2}$. In the opposite limiting case, when the profile $\Omega(r)$ is smooth and $l \gg l_\nu$, one can use the approximation of a perfect liquid and solve (7.12) taking into account small viscosity only near the singular points r_n of this equation where $\omega - n\Omega(r_n) = 0$. Analogously to the Rayleigh equation for a plane-parallel flow, the singular points r_n correspond to the phase resonance of the azimuthal wave with a flow in a vortex. The solution to (7.12) may be found by employing Lin's rule of path-tracing of singular points in a complex plane r. Note that the singular points exist in the solutions only for the harmonics $n \geq 2$ because for $n \leq 0$ the coefficient in (7.12) has no poles $\omega = n\Omega$ and, for $n = 1$, the solution has the form $\psi = r(\omega - \Omega)$ and has no singularities.

Multiplication of (7.12) by $inr\psi^*$ and integration by parts, with Lin's rule taken into consideration, yield

$$L(r_1) = -\pi n \left[\frac{|\psi|^2}{|d\Omega/dr|} \frac{dQ}{dr} \right]_{r=r_n} . \tag{7.44}$$

Comparison of this equation to (7.38) gives

$$(1 - |R_n|^2) \left[\frac{dQ}{dr} \right]_{r=r_n} . \tag{7.45}$$

The sign of the dissipation, i. e., amplification or attenuation of a cylindrical wave on reflection at a vortex, in a perfect liquid, is determined by the sign of the derivative of vorticity at the resonance point:

$$\frac{dQ}{dr} = r\frac{d^2\Omega}{dr^2} + \frac{d\Omega}{dr}.$$

The vorticity usually decreases monotonically with increasing distance from the vortex axis and, hence, the waves are amplified at reflection: $|R_n| > 1$.

The resonant amplification of IGWs at scattering by a cylindrical vortex in a perfect fluid occurs due to the kinetic energy of the particles synchronous to the wave, for which the wave frequency is multiple to their rotation frequency in the velocity field of the vortex. Here, one can clearly see the analogy with cyclotron damping (or amplification) of electromagnetic waves in a magnetized plasma (Zheleznyakov, 1977). However, unlike the homogeneous plasma where the frequency of rotation in a magnetic field is equal for all particles, the hydrodynamical flow has a continuous spectrum of rotation frequencies. The respective boundary value problem has the singularities that are determined by the poles $\omega = n\Omega(r)$ in the coefficients of the differential equations (7.12) or (7.7). A similar situation, namely, the velocity distribution of flow particles and resonance singular points in the differential equations for the amplitudes of small perturbations, was discussed in Sect. 2.1 in the frame of the plasma–hydrodynamic analogy for linear shear flows.

Consider now the magnitude of resonant amplification. The amplification of cylindrical harmonics is determined by (7.40). The magnitude of $\operatorname{Im} q_n$ in a perfect liquid depends on the profile of the relative angular velocity $\Omega(r)/\omega$ and can be found by solving (1.12) with the boundary conditions (7.40). The dependences $\operatorname{Im} q_n(\tilde{\Omega}/\omega)$ were calculated by Golemshtok and Fabrikant (1980) and Fabrikant (1990) for a vortex with the angular velocity profile

$$\Omega = \frac{\tilde{\Omega} a^2}{r^2}[1 - \exp(-r^2/a^2)]. \tag{7.46}$$

These dependences are plotted in Fig. 7.4 for the harmonics $n = 2, 3$. The magnitude of $\operatorname{Im} q_n$ decreases rapidly with increasing n.

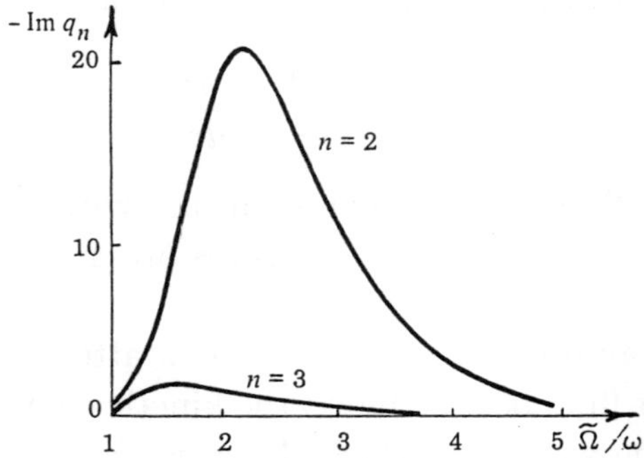

Figure 7.4: The $\operatorname{Im} q_n(\tilde{\Omega}/\omega)$ dependences for cylindrical waves on a vortex with continuous velocity profile.

The plane wave scattering at a vortex depends on the set of the reflection coefficients of cylindrical harmonics R_n for different n. By expanding the plane wave in cylindrical harmonics one can find the perturbation field far from the vortex in the form

$$p = e^{ikr\cos\varphi} + f(\varphi)\frac{e^{ikr}}{\sqrt{r}}$$

$$= \sum_n \frac{i^n e^{in\varphi}}{\sqrt{2\pi kr}} \left[e^{-i\left(kr-\frac{n\pi}{2}-\frac{\pi}{4}\right)} + (1+f_n\sqrt{2\pi ik})e^{i\left(kr-\frac{n\pi}{2}-\frac{\pi}{4}\right)}\right]. \quad (7.47)$$

Here, $f(\varphi) = \sum f_n e^{-in\varphi}$ is the sum amplitude of scattering and the partial amplitudes f_n are expressed through the reflection coefficients:

$$f_n = \frac{R_n - 1}{\sqrt{2\pi ik}}.$$

The cross-section of the plane wave absorption on a cylindrical vortex that describes the wave energy flux across the cylindrical surface around the vortex (at a unit energy flux density in an incident plane wave) can be readily found from (7.47)

$$\sigma_r = \frac{\tilde{c}}{\omega}\sum_{n\geq 2}(1-|R_n|^2). \quad (7.48)$$

For $\mu = \omega a/\tilde{c} \ll 1$, the cross-section of resonant absorption is determined by the harmonic $n = 2$:

$$\sigma_r = \frac{\tilde{c}}{\omega}(1-|R_2|^2) = \frac{\pi a}{8}\left(\frac{\omega a}{\tilde{c}}\right)^3 \operatorname{Im} q_2. \quad (7.49)$$

As was shown above, the cross-section of resonant absorption for a vortex in a perfect fluid may be negative. This effect is possible under the following conditions:

(i) the rotation frequency of particles in a vortex must exceed half the wave frequency so that the condition $\omega = 2\Omega(r) < 2\tilde{\Omega}$ should be met;

(ii) the flow vorticity must decrease from the vortex center to the periphery.

For the angular velocity profile (7.45), the effective magnitude of amplification characterized by the cross-section σ reaches its maximum at $\omega \approx \tilde{\Omega}/2$. The absorption cross-section in this case is $\sigma \sim \mu^3 a$, which is comparable to or even higher than the scattering cross-section of the waves (Fabrikant, 1982). Thus, it is highly probable that the effect of resonant amplification considered here may be observed in real conditions, for example, in the IGW propagation in an atmosphere or in the ocean in the presence of sufficiently intense localized vortices.

7.4 Scattering of long waves

The scattering cross-section of a plane wave on a vortex flow with the characteristic size a may be found by means of the method of matched asymptotic expansions in the parameter $\mu = \omega a/\tilde{c} \ll 1$. To this end, one needs to introduce the internal variable $\xi = r/a$ and the external variable $\sigma = \omega r/\tilde{c}$. Then, (7.37) gives the external expansion to an accuracy of μ^2:

$$\Phi_{e2} = A\left[\mathrm{H}_\alpha^{(1)}(\sigma) + \tilde{R}_n \mathrm{H}_\alpha^{(2)}(\sigma)\right] \underset{\sigma \gg n}{\longrightarrow} \sqrt{\frac{2}{\pi\sigma}}\left(\mathcal{E}_1 + \tilde{R}_n \mathcal{E}_2\right), \qquad (7.50)$$

where

$$\mathcal{E}_1 = \exp\left[-i\left(\sigma - \frac{n\pi}{2} - \frac{\pi}{4}\right) - \frac{i\pi}{2}\mu^2 \frac{n\Omega_0}{|n|\omega}\right],$$

$$\mathcal{E}_2 = \exp\left[i\left(\sigma - \frac{n\pi}{2} - \frac{\pi}{4}\right) + \frac{i\pi}{2}\mu^2 \frac{n\Omega_0}{|n|\omega}\right],$$

$\alpha^2 = n^2 + 2\mu^2 n\Omega_0/\omega$ and the constant Ω_0 is determined by the integral over the entire cross-section of the vortex core:

$$\Omega_0 a^2 = \frac{1}{2\pi}\int Q\, ds. \qquad (7.51)$$

The potential flow far from the vortex core has the form

$$\Omega(r \gg a) = \Omega_0 \frac{a^2}{r^2},$$

and the vorticity decreases exponentially as $Q\exp(-r^2/a^2)$. The reflection coefficient of the cylindrical wave is found from (7.39):

$$R_n = \tilde{R}_n \exp\left[i\pi\mu^2 \frac{n}{|n|}\frac{\Omega_0}{\omega}\right]. \tag{7.52}$$

The internal expansion, to an accuracy of μ^2, has the form (7.40). Matching the expansions (7.40) and (7.50) yields, accurate to μ^2, $\tilde{R}_n = 1$ for $|n| \geq 2$ and $\tilde{R}_n = 1 + i\pi q_n \mu^2/2$ for $n = \pm 1$.

The reflection coefficient R_0 for the harmonic $n = 0$ needs special consideration. By solving (7.7) at $\Omega_0 = 0$, $n = 0$ by the method of matched series expansions one can show that $|1 - R_0| \ll \mu^2$.

The exact solution to (7.12) for the harmonics $n = \pm 1$ has the form

$$\psi_{\pm 1} = r\left[\Omega(r) - \omega\right] \underset{r \gg a}{\longrightarrow} \frac{\Omega_0 a^2}{r} \mp \omega r. \tag{7.53}$$

Thus, one has $q_{\pm 1} = \mp \Omega_0/\omega$ for the harmonics $n = \pm 1$.

The differential scattering cross-section of a plane wave on a cylindrical vortex is expressed through the reflection coefficients of cylindrical harmonics:

$$d\sigma = |f(\varphi)|^2 d\varphi = \frac{\tilde{c}}{2\pi\omega}\left|\sum_{n=-\infty}^{n=+\infty}(R_n - 1)e^{-in\varphi}\right|^2 d\varphi. \tag{7.54}$$

The substitution of (7.52) for the reflection coefficients R_n yields

$$d\sigma = \frac{\pi a}{2}\left(\frac{a\Omega_0}{\tilde{c}}\right)^3 \frac{\omega}{\Omega_0}\cos^2\varphi \cot^2\frac{\varphi}{2} d\varphi. \tag{7.55}$$

This expression was found for the scattering cross-section of acoustic waves on a point vortex by Pitaevsky (1958) and Ferziger (1974) by means of a different technique, namely, the perturbation method to the Born approximation with respect to the parameter

$$\tilde{\mu} = \left(\frac{\omega a}{c}\right)^2 \frac{\Omega_0}{\omega}.$$

However, the validity of the perturbation method is questionable for the singular effective potential determined by a point vortex. The solution presented above may be considered as a justification of the correctness of the perturbation method for investigations of the wave scattering at point vortices.

It is noteworthy that, although the absorption cross-section (7.49) has at $\Omega_0 \sim \omega$ the same order of magnitude of $\mu^3 a$ as the scattering cross-section, the Born approximation is insufficient for its calculation. Instead, the next-order approximation of the perturbation theory is needed here. The expansion in the cylindrical functions presented above allows for simpler calculation of σ_r.

Chapter 8

Remote Wave Sensing of Moving Vortices

The interaction of waves with vortices is one of the most essential hydrodynamical processes that frequently determines the wave propagation in turbulent and other vortex flows. Among the different problems of the theory of wave propagation in vortex flows of great importance is investigation of the wave scattering at systems of localized vortices. In particular, acoustic scattering by vortices is vital for the diagnostics of vortex flows and control of turbulence.

The IGW scattering, absorption, and amplification at atmospheric cyclones, synoptic vortices in the ocean, convective cells, tornados, and the like are the fundamental mechanisms determining the wave dynamics in an atmosphere and in the ocean. Investigation of the characteristics of wave scattering at typical vortex formations occurring in real life gives an insight into the conditions of wave propagation in the media of interest and, on the other hand, allows for the diagnostics of vortex flows, which is often a cardinal applied problem.

The greatest majority of known vortex flows are determined by a relatively simple vorticity distribution. The velocity field in these cases may be much more complicated than the vorticity distribution because the vorticity is a conserved quantity and is transported by the particles of a perfect fluid without changes (Lamb, 1932; Lavrent'ev and Shabat, 1977). In particular, the vorticity is constant along the closed streamlines in a stationary flow. Consequently, it is more convenient to express the characteristics of the wave scattering at a vortex flow through the vorticity of this flow rather than through velocity field.

8.1 Acoustic and internal wave scattering at vortices in a stratified rotating medium

Consider a linear theory of scattering for acoustic-gravity waves as well as for IGWs of different scales (inertia-gravity waves including) at the quasistationary vortex flows with small Froude numbers (the definition will be given below) (Fabrikant, 1983; Stepanyants and Fabrikant, 1991). The field of the scattered waves is sought to the Born approximation of the perturbation theory. The amplitude of scattering, as will be shown a little later, is proportional to the Fourier component of vorticity, which greatly simplifies the investigation for a broad class of flows with concentrated vortices.

The consideration will be restricted to investigation of the plane horizontal motions the velocity of which has only two components $\mathbf{U}(x, y) = (U_x, U_y, 0)$. This class of flows includes a considerable part of the well-known vortex formations in stratified media. Consider the scattering of the harmonic waves with frequency ω at a quasistationary flow varying in the characteristic time $T \gg \omega^{-1}$. Equations (7.1) derived for small perturbations in the approximation of a plane-laminar rotating medium are taken as basic ones. By substituting (7.2) and regarding all the variables to be proportional to $e^{-i\omega t+iqz}$ one obtains for the amplitude of perturbations in a perfect fluid ($\nu = 0$) the following system of equations

$$\begin{aligned}
[-i\omega + (\mathbf{U}, \nabla)]\mathbf{v} + \nabla p + 2\,[\mathbf{\Omega}_0, \mathbf{v}] + (\mathbf{v}, \nabla)\mathbf{U} &= 0, \\
[-i\omega + (\mathbf{U}, \nabla)]\frac{p}{c_s^2} + \operatorname{div}\mathbf{v} + (iq - \Gamma)w &= 0, \qquad (8.1) \\
\{N^2 + [-i\omega + (\mathbf{U}, \nabla)]^2\}\,w + (iq + \Gamma)[-i\omega + (\mathbf{U}, \nabla)]p &= 0.
\end{aligned}$$

Here, the Brunt–Väisälä frequency N and the Ekkart parameter Γ are takeñ to be constant.

Note that the spatial homogeneity of the unperturbed medium in a horizontal plane is possible only for sufficiently weak flows with characteristic Froude number $\mathrm{Fr} = \Gamma U/N \ll 1$. This condition permits one to neglect the "funnel" effect, i. e., the deformation of the isopycnic surfaces under the action of the centrifugal force that pushes the liquid out of the vortex region. The corresponding change of the density at the center of the funnel is approximately $\Delta\rho \sim \rho\mathrm{Fr}$. The horizontal density inhomogeneity makes a correction to the scattered wave field that is proportional to $U\mathrm{Fr}$, i. e., to the squared flow velocity.

In order to employ the Born approximation, the ratio of the flow velocity to the phase velocity of the wave is taken as a small parameter. The system (8.1), with the incompressibility condition of the unperturbed two-dimensional vortex flow ($\mathrm{div}\,\mathbf{U} = 0$) taken into account, may give for the Fourier transform of the pressure field

$$P_{\mathbf{k}} = \iint\limits_{-\infty}^{+\infty} P(\mathbf{r}) \exp[-i(\mathbf{k},\mathbf{r})]\, d\mathbf{r} \tag{8.2}$$

the following integral equation

$$\frac{(k_0^2 - k^2)(N^2 - \omega^2)}{\omega^2 - 4\Omega_0^2} P_{\mathbf{k}} =$$

$$\frac{1}{2\pi} \iint P_{\mathbf{k}'}\, d\mathbf{k}' \left\{ \frac{(\mathbf{k}', \mathbf{U}_{\mathbf{k}-\mathbf{k}'})}{\omega} \left[\frac{N^2 - 3\omega^2}{\omega^2} + \Gamma^2 + q^2 + \frac{2\omega^2 k^2}{\omega^2 - 4\Omega^2} \right.\right.$$

$$\left. + \frac{\omega^2 (N^2 - \omega^2)}{(\omega^2 - 4\Omega_0^2)^2} \left(2(\mathbf{k},\mathbf{k}') - k'^2 - \frac{6i(\boldsymbol{\Omega}_0, [\mathbf{k}',\mathbf{k}])}{\omega} + \frac{4\Omega_0^2 (\mathbf{k},\mathbf{k}')}{\omega^2} \right) \right]$$

$$\left. + \frac{2i(N^2 - \omega^2)}{(\omega^2 - 4\Omega_0^2)^2} (\mathbf{U}_{\mathbf{k}-\mathbf{k}'}, [\boldsymbol{\Omega}_0, \mathbf{k}]) \left((\mathbf{k},\mathbf{k}') - k'^2 - \frac{2i(\boldsymbol{\Omega}_0, [\mathbf{k}',\mathbf{k}])}{\omega} \right) \right\}, \tag{8.3}$$

where $k_0^2 = (\omega^2 - 4\Omega_0^2) \left(\dfrac{1}{c_s^2} + \dfrac{\Gamma^2 + q^2}{N^2 - \omega^2} \right)$.

Let a plane wave

$$p(\mathbf{r}) = p_0 \exp[-i\omega t + i(\mathbf{k}_0\mathbf{r})], \qquad p_{\mathbf{k}} = (2\pi)^2 p_0 \delta(\mathbf{k} - \mathbf{k}_0)$$

with wavenumber $k_0 = |\mathbf{k}_0|$ be incident on the vortex flow. By substituting these expressions into the right-hand side of the integral equation (8.3) and expressing the flow velocity through its vorticity $\mathbf{Q} = \operatorname{rot} \mathbf{U}$ (from which follows $\mathbf{U}_{\mathbf{k}} = i[\mathbf{k}, \mathbf{Q}]/k^2$), one finds the pressure field in the scattered wave:

$$\frac{p}{p_0} = i \iint \frac{(\mathbf{Q}_{\mathbf{k}-\mathbf{k}_0}, [\mathbf{k}_0, \mathbf{k}])(\omega^2 - 4\Omega_0^2) I(\mathbf{k}_0, \mathbf{k}) e^{i(\mathbf{k},\mathbf{r})}}{4\pi^2 \omega (N^2 - \omega^2)(\mathbf{k} - \mathbf{k}_0)^2 (k^2 - k_0^2)} \, dk_{\mathrm{x}} dk_{\mathrm{y}}, \qquad (8.4)$$

where

$$I(\mathbf{k}_0, \mathbf{k}) = \frac{N^2 - 3\omega^2}{c_{\mathrm{s}}^2} + \Gamma^2 + q^2 + \frac{2\omega^2 k^2}{\omega^2 - 4\Omega_0^2} + \frac{\omega^2 (N^2 - \omega^2)}{(\omega^2 - 4\Omega_0^2)^2}$$

$$\times \left\{ 2(\mathbf{k}, \mathbf{k}_0) - k_0^2 - \frac{6i(\boldsymbol{\Omega}_0, [\mathbf{k}_0, \mathbf{k}])}{\omega} + \frac{4\Omega_0^2 (\mathbf{k}, \mathbf{k}_0)}{\omega^2} \right.$$

$$\left. + 2\left(\frac{\Omega_0}{\omega}\right)^2 [(\mathbf{k}, \mathbf{k}_0) - k^2] \left[1 + \frac{i\omega[(\mathbf{k}, \mathbf{k}_0) - k^2]}{2(\boldsymbol{\Omega}_0, [\mathbf{k}_0, \mathbf{k}])}\right] \right\}. \qquad (8.5)$$

Let us now calculate the amplitude of the plane wave scattering at the vortex sheet concentrated in the region with characteristic size L. Clearly, the scattered field is determined by the poles $k_x^2 + k_y^2 = k_0^2$ of the integrand in (8.4). The rule of path-tracing follows here from the causality principle. In particular, for $y > 0$, the integration contour in the complex halfplane $\operatorname{Re} k_y > 0$ goes round the pole from below, and for $y < 0$, from above. Note that (8.4) also contains another pole $k_y = k_{0y}$ that determines the quasistatic pressure perturbation field arising from the deformation of the vortex flow in the incident wave field. The corresponding solution $p(r)$ decays rapidly (r^{-n}, $n \geq 2$) with increasing distance to the region of size L that is occupied by the flow. Therefore, the contribution of this pole at $r \gg L$ may be neglected.

If the plane wave is incident at an angle φ_0 to the x-axis, i. e., $k_0 = k_0(\cos\varphi_0,\ \sin\varphi_0)$, then the scattered wave field has the form

$$\frac{p}{p_0} = -\frac{\omega^2 - 4\Omega_0^2}{N^2 - \omega^2} \int \frac{(\mathbf{Q}_{\mathbf{k}-\mathbf{k}_0}, [\mathbf{k}_0, \mathbf{k}])}{(\mathbf{k} - \mathbf{k}_0)^2} \frac{I(\mathbf{k}_0, \mathbf{k})}{4\pi\omega\eta} e^{i(k_x x + \eta y)}\, dk_x, \tag{8.6}$$

where $\mathbf{k}_0$ and $\mathbf{k} = (k_x, \eta)$ are the wave vectors of the incident and scattered waves, φ_0 and φ are the angles of incidence and scattering, respectively, and $\eta = \sqrt{k_0^2 - k_x^2}$. The asymptotic form of the scattered field in the far zone may be found by the method of stationary phase

$$\frac{p}{p_0} = f(\varphi, \varphi_0)\frac{e^{ikr}}{\sqrt{r}}, \tag{8.7}$$

where

$$f = \sqrt{\frac{k_0^3}{8\pi}} \cot\frac{\varphi - \varphi_0}{2} \Phi(\varphi - \varphi_0) \frac{Q_{\mathbf{k}-\mathbf{k}_0}}{\omega}, \qquad Q_{\mathbf{k}-\mathbf{k}_0} = |\mathbf{Q}_{\mathbf{k}-\mathbf{k}_0}|,$$

$$\Phi(\varphi) = \cos\varphi + \frac{\omega^2}{c_s^2 k^2} \frac{c_s^2 k^2 - \omega^2 + 4\Omega_0^2}{N^2 - \omega^2} \tag{8.8}$$

$$+ \frac{2i\omega\Omega_0}{\omega^2 - 4\Omega_0^2}\left[\mathrm{tg}\frac{\varphi}{2} - 2\sin\varphi - 2i(2\cos\varphi - 1)\frac{\Omega_0}{\omega}\right].$$

From this follows the expression for the amplitude of acoustic wave scattering at vortices in a homogeneous medium at $\omega \gg N$ (Fabrikant, 1983). For $\Omega_0 \ll \omega \ll N$, the amplitude of the IGW scattering is

$$f(\varphi - \varphi_0) = \sqrt{\frac{k_0^3}{8\pi i}} \frac{Q_{\mathbf{k}-\mathbf{k}_0}}{\omega} \cos(\varphi - \varphi_0) \cot\frac{\varphi - \varphi_0}{2}. \tag{8.9}$$

For the IGW moving at an angle to the horizon, this expression coincides with the expression for the horizontally propagating sound. Note that there is no backward scattering for acoustic and internal gravity waves; it is possible only for inertia-gravity waves.

If $Q_{k=0} \neq 0$, the amplitude of scattering has a singularity at $\varphi = 0$. As a consequence, the Born approximation is incorrect at small angles.

The singularity at small angles is involved with a rather slow decrease of the velocity field in the vortex which is proportional to r^{-1} as $r \to \infty$. Therefore, even the rays that are far from the vortex core refract into its field.

The vorticity $Q(r)$ may be expanded into multipole series

$$Q_k = Q_{k=0} + (\mathbf{k}, \mathbf{P}) + \dots, \tag{8.10}$$

where

$$Q_{k=0} = \int Q(\mathbf{r})\, d\mathbf{r} = 2\pi\kappa$$

is the total vorticity of the flow (κ being the velocity circulation around the vortex) and the vector

$$\mathbf{P} = \int \mathbf{r} Q(\mathbf{r})\, d\mathbf{r}$$

is the dipole moment of vorticity.

If the size of the vortex, L, is small compared to the wavelength ($kL \ll 1$), the consideration may be restricted to the first nonvanishing term of this series. Given the total flow vorticity $Q_{k=0} \neq 0$, the vortex flow can be replaced, to the first approximation, by the point vortex with circulation κ and the amplitude of scattering (8.9) then takes on the form found by Pitaevsky (1958) and O'Shea (1975) for a single point vortex:

$$f(\varphi - \varphi_0) = \frac{\pi\kappa}{c}\sqrt{\frac{i\omega}{2\pi\tilde{c}}}\cos(\varphi - \varphi_0)\cot\frac{\varphi - \varphi_0}{2}. \tag{8.11}$$

Thus, the long-wave scattering is determined by the oscillation of the vortex as a whole and does not depend on its internal structure. For the flows without circulation when $Q_{k=0} = 0$, the amplitude of scattering takes the form

$$f(\varphi) = \sqrt{\frac{k_0^5}{8\pi i}}\frac{\Phi(\varphi - \varphi_0)}{\omega}\cos(\varphi - \varphi_0)$$

$$\times\left[P_{\mathrm{y}}(\cos\varphi + \cos\varphi_0) - P_{\mathrm{x}}(\sin\varphi + \sin\varphi_0)\right]. \tag{8.12}$$

The function $f(\varphi)$ is bounded in this case at any angles. The quadripole and the subsequent terms of the expansion of the amplitude of scattering in the parameter kL are calculated analogously.

Thus far, the vorticity of the unperturbed flow $Q(\mathbf{r})$ was regarded as an arbitrary function of the two-dimensional radius-vector $\mathbf{r}$ rapidly decreasing at $r = |\mathbf{r}| \to \infty$ so that the Fourier transform of $Q(\mathbf{k})$ was finite. Consider now the axially symmetric flow whose Fourier transform depends only on $k = |\mathbf{k}|$. Let us take for the calculation of $Q(k)$ a polar system of coordinates and orient the polar axis along k. This gives

$$Q(k) = \int_0^\infty \int_0^{2\pi} rQ(r)e^{ikr\cos\psi}\, d\psi dr = 2\pi \int_0^\infty r\mathrm{J}_0(kr)Q(r)\, dr. \tag{8.13}$$

This integral converges if, at $r \to \infty$, $Q(r)$ decreases in proportion to r^{-s} where $s > 3/2$.

For the Ozeen vortex one has the vorticity and its spectrum:

$$Q(r) = 2B\exp(-r^2/L^2), \qquad Q(k) = 2\pi BR^2\exp(-k^2L^2).$$

Then, for the amplitude of scattering (8.7) one gets

$$f = \frac{\pi BL^2}{\omega}\sqrt{\frac{k^3}{2\pi i}}\Phi(\varphi-\varphi_0)\cot\frac{\varphi-\varphi_0}{2}\exp\left[-4k^2L^2\sin^2\frac{\varphi-\varphi_0}{2}\right]. \tag{8.14}$$

Let us find the amplitude of long-wave scattering ($kL \ll 1$) at a vortex with decaying vorticity $Q = BF(r^2/L^2)$. If the circulation

$$C \equiv \int_0^\infty F(\xi)\, d\xi \neq 0,$$

then, as was shown above, such a flow may be approximated by a point vortex. For the flow without circulation ($C = 0$) one has

$$f = \frac{k_0^3L^4B}{4\omega}\sqrt{\frac{i\pi k_0}{2}}\sin(\varphi-\varphi_0)\Phi(\varphi-\varphi_0)\int_0^\infty \xi F(\xi)\, d\xi. \tag{8.15}$$

The amplitude of scattering f in this case is bounded at any angles and the total IGW scattering cross-section

$$\sigma \sim \left(\frac{Q}{\omega}\right)^2 \frac{\pi^2 k_0^7 L^8}{128}.$$

Note that the corresponding cross-section for the vortices with dipole moment $P \sim QL$ has the order of magnitude

$$\sigma \sim \left(\frac{Q}{\omega}\right)^2 \frac{k_0^5 L^6}{32\pi},$$

thus exceeding the cross-section of the axially symmetric vortex by a factor of $(k_0 L)^{-2}$.

8.2 Wave scattering at a system of point vortices

The modelling of a flow by a system of point vortices that is broadly employed in hydrodynamics (Batchelor, 1970) gives simple analytical expressions for the amplitude of scattering. It is natural to draw an analogy to the point vortices with point charges in electrodynamics (Landau and Lifshits, 1973). There is, however, a notable difference which lies in that the sound scattering depends not only on its interaction with a vortex core but also on the refraction in the velocity field far beyond the core, which leads to the singularity in the amplitude of scattering at small angles. Whereas in the case of the electromagnetic wave scattering at a point vortex the Coulomb field of the charge does not affect the wave field by virtue of the superposition principle.

The flow vorticity may be approximated by

$$Q(\mathbf{r}) = \sum_n 2\pi\kappa_n \delta(\mathbf{r} - \mathbf{r}_n), \tag{8.16}$$

where κ_n is the circulation of the n-th vortex. The amplitude of scattering in such a flow is written as

$$f = \frac{\pi}{\tilde{c}}\sqrt{\frac{i\omega}{2\pi\tilde{c}}}\Phi(\varphi - \varphi_0)\mathrm{ctg}\frac{\varphi - \varphi_0}{2}\sum_{\mathrm{n}} \kappa_{\mathrm{n}} \exp\left[i(\mathbf{k} - \mathbf{k}_0)\mathbf{r}_{\mathrm{n}}\right]. \tag{8.17}$$

The scattering cross-section averaged over the time that is larger than the characteristic period T of the vortex motion has the form

$$d\sigma = \langle |f(\varphi)|^2 \rangle d\varphi$$

$$= \frac{\pi\omega}{2\tilde{c}^2}\Phi^2(\varphi-\varphi_0)\cot^2\frac{\varphi-\varphi_0}{2}\left\langle\left|\sum_n \kappa_n \exp[i(\mathbf{k}-\mathbf{k}_0)\mathbf{r}_n]\right|^2\right\rangle d\varphi, \quad (8.18)$$

where the angular brackets mean averaging over time. The long-wave scattering ($kL \ll 1$) at $\kappa = \sum \kappa_n \neq 0$ is described by the scattering amplitude (8.11) and the scattering cross-section

$$d\sigma = \frac{\pi\omega}{2\tilde{c}^2}\left|\sum_n \kappa_n\right|^2 \Phi^2(\varphi-\varphi_0)\cot^2\frac{\varphi-\varphi_0}{2}\,d\varphi. \quad (8.19)$$

For short waves ($kL \gg 1$), the exponents in (8.18) are rapidly oscillating functions of time because the i-th vortex changes its position by distance L in time T. Consequently, the averaging yields

$$d\sigma = \frac{\pi\omega}{2\tilde{c}^2}\left(\sum_n \kappa_n^2\right)\Phi^2(\varphi-\varphi_0)\cot^2\frac{\varphi-\varphi_0}{2}\,d\varphi. \quad (8.20)$$

It should be noted that this expression does not hold at small angles of scattering $|\varphi-\varphi_0| \leq (kL)^{-1}$ when $|(\mathbf{k}-\mathbf{k}_0, \mathbf{r}_i-\mathbf{r}_j)| \leq 1$.

Let us now find the value of coherent scattering, i. e., the scattering at the frequency of the incident wave (Landau and Lifshits, 1973). The scattered wave field at the incident wave frequency is determined by the expression $\langle f(\varphi)\rangle$. As a consequence, the cross-section of coherent scattering has the following form

$$d\sigma_c = |\langle f(\varphi)\rangle|^2 d\varphi. \quad (8.21)$$

For $kL \ll 1$ and $\sum \kappa_n \neq 0$ the exponent $\exp[i(\mathbf{k}-\mathbf{k}_0)\mathbf{r}_n]$ may be taken to be approximately equal to unity. This gives the amplitude of scattering (8.11). Comparison with (8.19) yields $d\sigma_c = d\sigma$, i. e., the long-wave scattering is completely coherent in this case. Whereas for $kL \gg 1$ all the terms of the sum in (8.18) are rapidly oscillating functions and vanish on averaging; thus, the scattering of short waves is completely incoherent.

Finally, let us calculate the amplitude of scattering at some simple systems of point vortices. Consider, in particular, a "double" vortex (Fig. 8.1a), i. e., a pair of vortices with identical circulation κ which are rotating with frequency $\Omega_0 = 2\kappa/d^2$ relative to one another (here d is the distance between vortices).

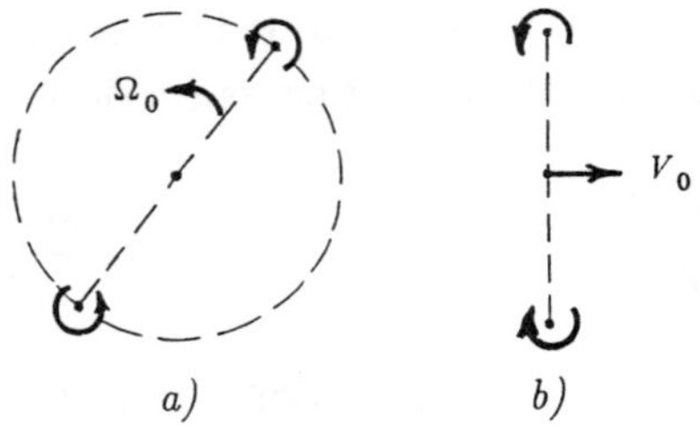

Figure 8.1: Systems of point vortices:
(a) "double" vortex, (b) vortex pair.

The expressions for the coordinates of these vortices are written as

$$\mathbf{r}_i = \frac{d}{2}\left(\cos(\Omega_0 t + i\pi),\ \cos(\Omega_0 t + i\pi)\right), \quad i = 1, 2. \tag{8.22}$$

Then, the amplitude of scattering is

$$f(\varphi) = \frac{2\pi\kappa}{\tilde{c}}\sqrt{\frac{i\omega}{2\pi\tilde{c}}}\Phi(\varphi - \varphi_0)\cot\frac{\varphi - \varphi_0}{2}$$

$$\times \cos\left\{\frac{\omega d}{2c}\left[(\cos\varphi - \cos\varphi_0)\cos\Omega_0 t + (\sin\varphi - \sin\varphi_0)\sin\Omega_0 t\right]\right\}$$

$$= \frac{2\pi\kappa}{\tilde{c}}\sqrt{\frac{i\omega}{2\pi\tilde{c}}}\Phi(\varphi - \varphi_0)\cot\frac{\varphi - \varphi_0}{2}$$

$$\times \sum_{\mathrm{m}}(-1)^m \mathrm{J}_m\left(\frac{\omega d}{\tilde{c}}\sin\frac{\varphi - \varphi_0}{2}\right)\cos\left[m\left(\Omega_0 t - \frac{\varphi + \varphi_0}{2}\right)\right]. \tag{8.23}$$

Clearly, the wave field scattered at such an oscillating flow contains an infinite spectrum of combination harmonics with frequencies $\omega \pm m\Omega_0$ $(m = 0, 1, 2, \dots)$.

For a pair of the vortices with opposite circulations (Fig. 8.1b) moving along the x-axis with velocity $V = \kappa/d$ one has

$$\mathbf{r}_i = \left(V_0 t,\ (-1)^i d/2\right), \quad i = 1, 2. \tag{8.24}$$

The amplitude of scattering is written as

$$f(\varphi) = \frac{2\pi\kappa}{\tilde{c}}\sqrt{\frac{-i\omega}{2\pi\tilde{c}}}\Phi(\varphi - \varphi_0)\cot\frac{\varphi - \varphi_0}{2}$$

$$\times \sin\left[\frac{\omega d}{2c}(\sin\varphi - \sin\varphi_0)\right]\exp\left[\frac{i\omega V}{\tilde{c}}(\cos\varphi - \cos\varphi_0)t\right]. \tag{8.25}$$

The flow is stationary in the reference system moving with the vortex pair with velocity V and the frequency of the scattered wave is equal to the incident wave frequency which, for the moving observer, is expressed by (Blokhintsev, 1981):

$$\omega' = \omega\left(1 - \frac{V}{\tilde{c}}\cos\varphi_0\right). \tag{8.26}$$

Therefore, in the motionless reference system, the frequency of the wave scattered at an angle φ differs from the incident wave frequency by the magnitude of the Dopler shift $\Delta\omega = \omega(V/\tilde{c})(\cos\varphi - \cos\varphi_0)$ to the first approximation with respect to the effective Mach number $\mathrm{M} = V/\tilde{c}$ (Tanaka and Ishii, 1982; Fabrikant, 1983).

8.3 Remote acoustic sensing of moving vortices

Knowledge of the characteristic features of the scattered signal such as Dopler frequency shift, the presence of combination harmonics, angular directivity pattern and their dependence on the wave frequency allows, in principle, for remote diagnostics of a vortex flow which gives such characteristics as the velocity of vortex motion and the characteristic distance between vortices, the frequencies of intrinsic oscillations, integral intensity, and others. The use of acoustic waves appears quite promising for this purpose (Ferziger, 1974). It is noteworthy that the linear acoustic scattering at given vortices is the most favorable effect for acoustic diagnostics because, unlike nonlinear phenomena, it does not need very high intensity of sound and is not so sensitive to fine details of the flow as, for example, the effects of resonant damping or amplification that were considered above.

Remote acoustic sensing is broadly employed for the diagnostics of vortex turbulence in an atmosphere (Tatarsky, 1967; McAllister *et al.*, 1969; Krasnenko, 1986). The method of acoustic sensing of flows has a number of advantages as compared to the contact methods or to the sensing by electromagnetic waves (Little, 1969; Derr and Little, 1970; Russel *et al.*, 1974). Acoustic sensing gives information on a flow in short time, for large volumes, and in almost inaccessible places. At the same time, acoustic methods do not introduce into the flow structure any considerable perturbing effects which are inevitable with the use of contact methods. The theoretical and experimental results demonstrating the potentialities of acoustic sensing on an example of laboratory vortex flows were obtained by Gromov *et al.* (1972) and will be considered below.

8.3.1 Periodic vortex wake

A periodic vortex wake formed as a result of periodic vortex shedding behind a bluff body is an example of vortex flows that is highly significant in practical life and extremely interesting for theoreticians. In spite of a great amount of papers devoted to such flows (Chen, 1975), they are still attracting the attention of the researchers interested in the problems such as the resistance of the bodies moving in a flow (Kochin *et al.*, 1963), aerodynamic sound generation (Blokhintsev, 1981), aerohydroelasticity (Vol'mir, 1976, 1979), and others.

Consider the scattering of high-frequency acoustic waves at a periodic vortex wake behind a cylinder at small Mach numbers of the incident flow and analyse the potentialities of remote acoustic sensing for different characteristics of vortex wakes. A theoretical model is based on the approximation of the wake by a double chain of point vortices (Kármán vortex street, Fig.8.2) with vorticity in the form

$$Q(\mathbf{r}) = 2\pi\kappa \sum_{n=-\infty}^{n=+\infty} \left[\delta\left(x - \frac{l}{4} - nl + U_0 t\right) \delta\left(y - \frac{h}{2}\right) - \delta\left(x + \frac{l}{4} - nl + U_0 t\right) \delta\left(y + \frac{h}{2}\right) \right]. \quad (8.27)$$

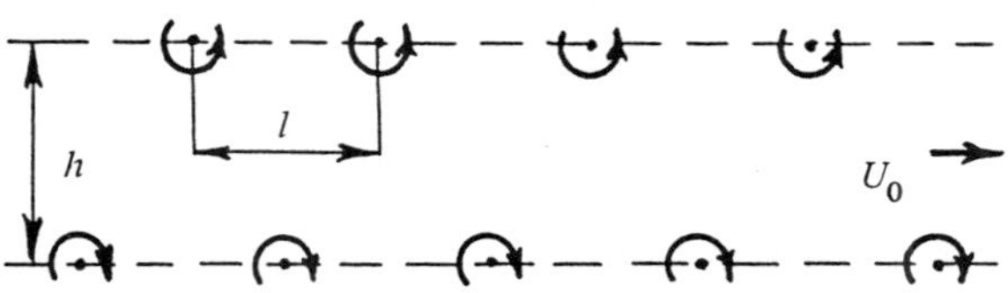

Figure 8.2: Kármán vortex street.

The point vortices are arranged in chess-board order and move in the positive direction of the x-axis with velocity U_0 that is specified by vortex circulation and by the velocity U_∞ with which the flow is incident on the cylinder:

$$U_0 = U_\infty - \frac{\pi\kappa}{l\sqrt{2}}. \quad (8.28)$$

It is the only possible configuration of the double vortex chain that is stable in the linear approximation (Kochin *et al.*, 1973). The ratio of the wake's width h to its period l is determined from the equality

$$\alpha = \pi\frac{h}{l} = \ln(1 + \sqrt{2}) \approx 0.881. \quad (8.29)$$

The flow around such a system of point vortices has a complex velocity

$$W \equiv -U_x + iU_y = \frac{i\pi\kappa}{l} \left\{ \cot\left[\frac{1}{l}(x - U_0 t - l/4 + iy - ih/2)\right] \right.$$

$$- \cot \left[\frac{1}{l}(x - U_0 t + l/4 + iy + ih/2) \right] \Big\} - U_\infty. \qquad (8.30)$$

This readily yields the expressions for the mean velocity along the axis of the wake ($y = 0$):

$$\langle U \rangle \equiv \frac{1}{l} \int_0^l U_x(y=0)\, dx = U_\infty - \frac{2\pi\kappa}{l} \qquad (8.31)$$

and the root-mean-square oscillatory velocity with respect to the mean flow on the wake's axis:

$$\tilde{U} \equiv \sqrt{\langle U^2 \rangle - \langle U \rangle^2} = \left[\frac{1}{l} \int_0^l (U_x^2 + U_y^2)_{|y=0}\, dx - \langle U \rangle^2 \right]^{1/2}$$

$$= \frac{2\pi\kappa}{l} \sqrt{\sqrt{2} - 1} \approx 0.64 \frac{2\pi\kappa}{l}. \qquad (8.32)$$

The parameters of Kármán vortex street are determined by the dependence of the dimensionless parameters $St = f_0 d/U_\infty$ (where $f_0 = U_0/l$ is the frequency of vortex shedding), l/d, $\Gamma = 2\pi\kappa/lU_\infty$, and others on the Reynolds number $\mathrm{Re} = U_\infty d/\nu$. These dependences were found experimentally as well as by means of semiempiric calculations (Chen, 1975). Note that the parameters of the wake depend relatively weakly on small perturbations: cylinder roughness, turbulence in an incident flow, etc.

The field of the scattered sound, when the plane wave in a plane wake is incident at an angle $\psi = \pi/2 - \varphi_0$ to the normal of the wake's axis, is described by (8.6), given $N = \Omega_0 = \Gamma = q = 0$. Let us substitute into this expression the Fourier transform of vorticity

$$Q(\mathbf{k}) = \frac{8\pi^2 i\kappa}{l} \sum_{\mathrm{m}=-\infty}^{\mathrm{m}=+\infty} \exp\left(\frac{2\pi i m}{l} U_0 t \right) \sin \frac{\pi m + k_{\mathrm{y}} h}{2} \delta\left(k_{\mathrm{x}} - \frac{2\pi m}{l} \right)$$

$$(8.33)$$

found from (8.31) using

$$\sum_{n=-\infty}^{n=+\infty} e^{inz} = 2\pi \sum_{n=-\infty}^{n=+\infty} \delta(z - 2\pi m). \qquad (8.34)$$

From the expressions given above one can find the scattered acoustic field

$$\frac{p}{p_0} = -i\mathrm{M} \sum_{\mathrm{m}=-\infty}^{\mathrm{m}=+\infty} \begin{cases} T_{\mathrm{m}} \exp(-i\Delta\omega_{\mathrm{m}} t - ik_{\mathrm{mx}} x - ik_{\mathrm{my}} y), & y > 0, \\ R_{\mathrm{m}} \exp(-i\Delta\omega_{\mathrm{m}} t - ik_{\mathrm{mx}} x + ik_{\mathrm{my}} y), & y < 0. \end{cases} \tag{8.35}$$

Here,

$$k_{mx} = \frac{\omega}{c_s} \sin\psi_m = \frac{\omega}{c_s}(\sin\psi_0 + ms), \quad k_{my} = \frac{\omega}{c_s}\cos\psi_m,$$

$$\mathrm{M} = \frac{2\pi\kappa}{lc_s}, \quad s = \frac{\lambda}{l} = \frac{2\pi c_s}{\omega l}, \quad \Delta\omega_m = \frac{2\pi m U_0}{l} = 2\pi m f_0.$$

The refraction and reflection coefficients T_m and R_m are written, respectively, as

$$T_m = \frac{\sin[2(\psi_m - \psi_0)]\sin[\pi m/2 + (\omega h/2c_s)(\cos\psi_m - \cos\psi_0)]}{\cos\psi_m[(2\pi m c_s/\omega l)^2 + (\cos\psi_m - \cos\psi_0)^2]},$$

$$R_m = \frac{\sin[2(\psi_m - \psi_0)]\sin[-\pi m/2 + (\omega h/2c_s)(\cos\psi_m + \cos\psi_0)]}{\cos\psi_m[(2\pi m c_s/\omega l)^2 + (\cos\psi_m + \cos\psi_0)^2]}. \tag{8.36}$$

The periodic system of vortices is a diffraction lattice, moving with velocity U_0, at which a plane acoustic wave is scattered. Thus, the scattered acoustic field is a sum of refracted and reflected harmonics. The harmonics with larger n decrease exponentially at $|y| \to \infty$. Only a finite number of harmonics that satisfy the condition

$$|\sin\psi_m| = |\sin\psi_0 + ms| \leq 1 \tag{8.37}$$

are plane waves propagating at angles ψ_m. Such solutions are typical of the wave scattering at periodic diffraction lattices (Golyamin, 1979). When scattering occurs at a moving lattice, the Dopler effect gives rise to the frequency shift of the diffracted sound: the frequency of the m-th harmonic differs from the frequency of the incident sound by $\Delta\omega_m$.

In a particular case of normal incidence ($\psi_0 = 0$), the harmonics propagate at angles $\psi_m = \arcsin(ms)$ and the coefficients of refraction and reflection have the form (Fig. 8.3):

$$T_m = \frac{ms\sin[(\alpha/s)(\cos\psi_m - 1) + m\pi/2]}{\cos\psi_m - 1},$$
$$R_m = \frac{ms\sin[(\alpha/s)(\cos\psi_m + 1) - m\pi/2]}{\cos\psi_m + 1}. \tag{8.38}$$

The parameter α is determined in the formula (8.29).

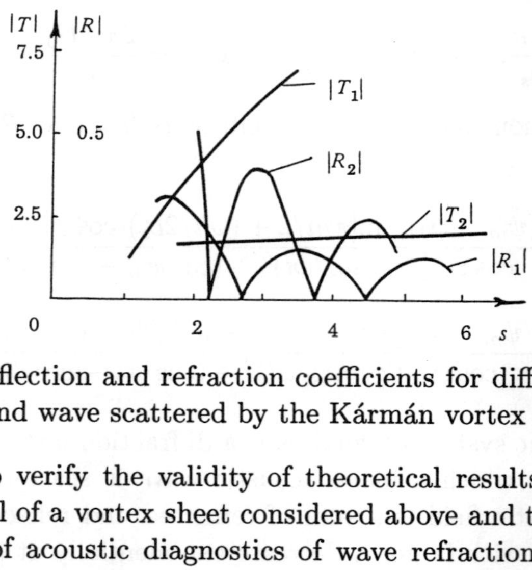

Figure 8.3: Reflection and refraction coefficients for diffraction harmonics of the sound wave scattered by the Kármán vortex street.

In order to verify the validity of theoretical results based on the idealized model of a vortex sheet considered above and to elucidate the potentialities of acoustic diagnostics of wave refraction on a periodic vortex lattice, a series of experiments on the supersonic scattering at a wake behind a streamlined body were carried out by Gromov *et al.* (1982,b). The experimental setup is presented in Fig. 8.4.

The cylinder was a rigid polished rod of diameter $d = 0.15$ cm. The level of turbulence in the incident air flow did not exceed 6% of the average velocity $U_\infty = 1.6$ m/s. Two hot-wire anemometers spaced apart were used to measure the root-mean-square oscillatory velocity at the wake's axis, $\tilde{U}$ (related to the vortex circulation by (8.32)), the frequency of vortex shedding, and the spatial period of the vortex street.

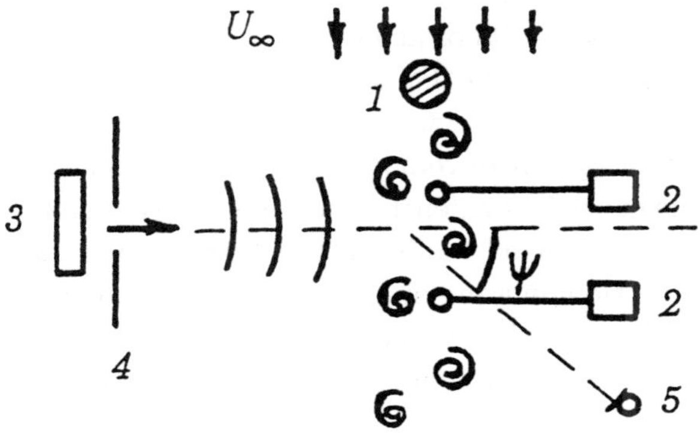

Figure 8.4: Scheme of experiment on ultrasound scattering by a vortex wake: 1 – cylinder, 2 – hot-wire anemometers, 3 – acoustic radiator, 4 – screen, 5 – microphone.

The entire region of the periodic vortex wake was in the Fraunhofer zone of acoustic radiator with frequency $f = 100 \div 200$ kHz. The sound wave incident on the wake is nearly plane under these conditions. The incident and scattered signals were measured by the Dopler frequency shift that is caused by the vortex motion. This enabled the researchers to isolate the weak quasimonochromatic scattered signal against the background of acoustic noises and to determine its parameters.

Remote acoustic sensing of the parameters of the wake was performed by the characteristics of the refraction coefficient at the first harmonic $m = 1$. The scattered waves had a quasidiscrete spectrum with a well pronounced spectral maximum shifted relative to the frequency of the scattered signal by $\Delta f = \Delta\omega_1/2\pi$ that corresponds to the frequency of vortex shedding. The angular directivity pattern of the amplitude of the harmonic $m = 1$ was measured by means of a microphone placed rather far from the wake. The directivity pattern had a well pronounced maximum at $\psi = \psi_1$ which indicated the presence of spatial periodicity in the wake. The average value of $l = c_s/(f \sin\psi_1)$ found from these data must coincide with the spatial period of the wake. Finally, the intensity of the scattered harmonic permits one to determine the value of vortex circulation by (8.35).

Table I compares results of acoustic sensing of the wake parameters and data obtained by direct calculations. One can see a good agreement between results of measurements of time and space periods of the flow. Less accurate is the result of remote sensing of the magnitude of vortex circulation. The cause is, evidently, the difference of the real vortex flow from the strongly idealized model of the Kármán vortex street. However, sound diffraction on a moving periodic system of vortices proved to be a rather rough effect that weakly depends on the defects of a vortex diffraction lattice. Therefore, in spite of a rather rough approximation of the flow, the wake's parameters found by the measurements of the scattered sound demonstrate a quite satisfactory agreement with results of direct measurements.

Table 1

	I	II	III
f_0, Hz	190 ± 30	165 ± 10	175 ± 15
$\mathrm{St} = \dfrac{fd}{U_0}$	0.18 ± 0.02	0.16 ± 0.02	0.17 ± 0.03
l, cm	0.65 ± 0.15	0.65 ± 0.05	0.52 ± 0.05
l/d	$4.5 \pm 1,0$	3.7 ± 0.4	3.5 ± 0.3
$\Gamma = \dfrac{2\pi\kappa}{lU_\infty}$	0.4 ± 0.05	0.45 ± 0.1	0.33 ± 0.1
$\dfrac{2\pi\kappa}{lU_\infty}$, cm	64 ± 10	69 ± 15	53 ± 15

In column I of this Table results of calculations within the framework of semiempirical theory and experimental data by Chen (1975) are presented. Data by Gromov *et al.* (1982b) obtained by direct mesuarements by means of hot-wire anemometers are given in the column II.

Results of calculations based on the data of mesuarements of scattered sound are presented in the column III.

Thus, results of the experiments performed by Gromov *et al.* (1982b) confirm that acoustic sensing of vortex wakes *is* possible. It is also verified by the results obtained by Rudnitsky (1989, 1990) who investigated in ample detail the technique of acoustic sensing of a vortex wake and other vortex flows making allowances for the turbulence present in the flow. Measurements of statistical characteristics of the sound signal scattered at the turbulent current gives additional information about the turbulent flow. It is worth noting that the remote acoustic sensing may be employed for the coherent structures in shear flows which are difficult to be measured by contact methods (Ezersky *et al.*, 1983).

8.3.2 Vortex rings

Investigation of vortex rings that was initiated back in the last century (Lamb, 1932) is still attracting great attention of researchers. Quite a great number of the effects important in real life that are determined by such vortices are known today. These include vortex transport of impurity and heat (Lavrent'ev and Shabat, 1977), excitation of internal and surface waves in a stratified liquid (Boyarintsev *et al.*, 1982; Dahm *et al.*, 1989), vortex sound generation (Klyatskin, 1966), and others.

In some instances (e. g., for vortices appearing at powerful explosions, for modulation of the rate of submerged jetc, etc.) one needs to locate the available intense vortices and determine their parameters by means of remote sensing. Such diagnostics may be performed by employing acoustic location (Ferziger, 1974; Howe, 1983). Below the reader will find results of investigations of acoustic scattering at vortex rings (Gromov *et al.*, 1982a) which enable one to find the parameters accessible to remote measurements.

Consider a free three-dimensional flow whose vorticity $\mathbf{\Omega} = \mathrm{rot}\,\mathbf{U}$ is localized in the finite region of characteristic size L. If a plane monochromatic wave is incident on the flow, then in the first approximation to the Mach number $\mathrm{M} = U/c_s$ one can obtain

$$\frac{\rho_1}{\rho_0} = \frac{2i(\mathbf{k}_0, \mathbf{k})([\mathbf{k}, \mathbf{k}_0], \boldsymbol{\Omega}_{\mathbf{k}-\mathbf{k}_0})}{\omega(k^2 - \omega^2/c_s^2)(\mathbf{k}-\mathbf{k}_0)^2} e^{i(\mathbf{k},\mathbf{r})}\, d\mathbf{k}, \tag{8.39}$$

where ρ_0 and ρ_1 are the amplitudes of density perturbations in the incident and scattered waves. Analogously to the procedure described in Sect. 8.1, one can find the scattered field in the far zone that is expressed through the amplitude of scattering

$$\frac{\rho_1}{\rho_0} = \frac{i\omega}{2\pi c_s^2} \frac{(\mathbf{n}_0, \mathbf{n})([\mathbf{n}, \mathbf{n}_0], \boldsymbol{\Omega}_{\mathbf{k}_r - \mathbf{k}_0})}{(\mathbf{n}-\mathbf{n}_0)^2} \frac{\exp(i\omega r/c_s)}{r}. \tag{8.40}$$

Here, $\mathbf{n} = \mathbf{r}/|\mathbf{r}|$, $\mathbf{n}_0 = \mathbf{k}_0/|\mathbf{k}_0|$, and $\mathbf{k}_r = \mathbf{n}\omega/c_s$.

Consider a particular case of a flow with axial symmetry. Let us orient the coordinate axis (Fig. 8.5) so that $\mathbf{k}_0 = k_0(\sin\theta_0,\ \cos\theta_0,\ 0)$, $\boldsymbol{\Omega} = \Omega_0(y, \mathbf{r}_\perp)\mathbf{n}_\perp$, where $\mathbf{r}_\perp = (x, 0, z)$, $\mathbf{n}_\perp = [\mathbf{n}_y, \mathbf{r}_\perp]/r_\perp$.

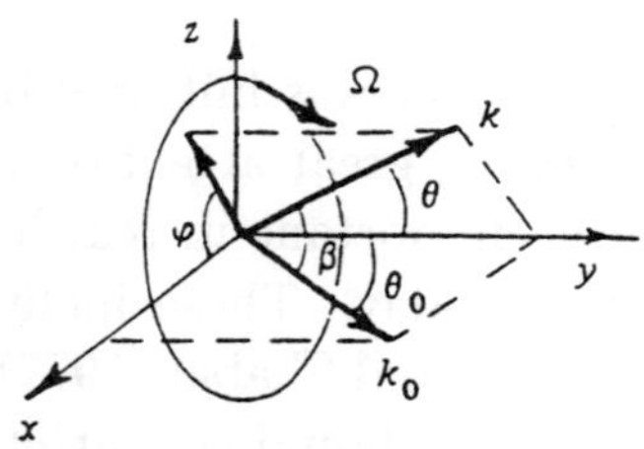

Figure 8.5: Sound scattering by a vortex ring.

In this case, the amplitude of scattering may be presented in the following form

$$f(n, n_0) = -\frac{\omega}{2c_s^2} \frac{\cos\beta(\cos\theta + \cos\theta_0)}{q_\perp}$$

$$\times \int_0^\infty r_\perp \mathrm{J}_1\left(\frac{\omega r_\perp}{c_s} q_\perp\right) \Omega_0(k_y - k_{0y}, r_\perp)\, dr_\perp, \tag{8.41}$$

where $\cos\beta \equiv (\mathbf{n}, \mathbf{n}_0) = \cos\theta\cos\theta_0 + \sin\theta\sin\theta_0\cos\varphi$, and

$$q_\perp = \sqrt{(\sin\theta\cos\varphi - \sin\theta_0)^2 + \sin^2\theta\sin^2\varphi},$$

θ and φ being the angles determining vector $\mathbf{n}$ in the spherical system of coordinates. Note that if the incident wave is propagating normally to the vortex axis ($\cos\theta_0 = 0$), the forward scattering is zero: $f(\mathbf{n}, \mathbf{n}_0) = 0$. For the simplest vortex ring, i. e., an infinitely thin vortex sheet closed in a ring, one has

$$\Omega_0(y, r_\perp) = \Gamma\delta[y - y_0(t)]\delta(r_\perp - r). \tag{8.42}$$

Then, the amplitude of scattering has the form

$$f(\mathbf{n}, \mathbf{n}_0) = \frac{-\Gamma\omega L}{2c_s^2 q_\perp}\cos\beta(\cos\theta + \cos\theta_0)\mathrm{J}_1\left(\frac{\omega L q_\perp}{c_s}\right)\exp\left(\frac{i\omega}{c_s}y_0(t)q_y\right), \tag{8.43}$$

where $q_y = \cos\theta - \cos\theta_0$. Note that calculation of the directivity diagram for the scattered sound by (8.43) gives the results which coincide with the numerical calculations performed by Kambe and Oo (1981).

At small angles of scattering ($\beta \to 0$), (8.43) yields

$$f(\mathbf{n}, \mathbf{n}_0) = -\frac{\Gamma}{2c_s}\left(\frac{\omega L}{c_s}\right)^2\cos\theta. \tag{8.44}$$

If the vortex moves with constant velocity U_0 = const, then the Dopler frequency shift of the scattered sound in the far zone is

$$\Delta\omega = \frac{\omega U_0}{c_s}(\cos\theta - \cos\theta_0)$$

(we remind the reader that the flows with small Mach numbers are considered).

Let us investigate the scattering at the vortex ring formed by the region of vorticity that can be represented as a torus. Let the radius of the torus cross-section be $a \ll L$. Such vortices are a good model for the ring vortices observed in experiments. If the vorticity is constant inside the torus and the circulation around its cross-section is equal to Γ, then the amplitude of scattering (8.41) takes the form

$$f(\mathbf{n}, \mathbf{n}_0) = -\frac{\Gamma\omega L}{2c_s^2 q_\perp}F\cos\beta(\cos\theta + \cos\theta_0)\exp\left(\frac{i\omega}{c_s}y_0(t)q_y\right), \tag{8.45}$$

where

$$F = \frac{2L}{\pi a} \int_{-\pi/2}^{\pi/2} \left(1 + \frac{a}{L}\sin\gamma\right) \frac{\sin\left[(\omega a/c_s)q_y \cos\gamma\right]}{\omega q_y/c_s} \mathrm{J}_1(X)\cos\gamma\, d\gamma, \quad (8.46)$$

$$X = \frac{\omega q_\perp}{c_s}(L + a\sin\gamma).$$

This expression may also be obtained by means of an alternative procedure, i. e., by dividing the torus into vortex sheets and summing the contributions into (8.43) of the individual sheets. Under the condition $a/L \ll 1$, one can obtain an approximate expression

$$F = \frac{2\mathrm{J}_1(\omega a q_y/c_s)}{\pi\omega a q_y/c_s} \int_{-\pi/2}^{\pi/2} \mathrm{J}_1\left(\frac{\omega q_\perp}{c_s}(L + a\sin\gamma)\right) d\gamma. \quad (8.47)$$

Comparison of (8.45) and (8.47) with (8.43) for an infinitely thin vortex sheet demonstrates that allowance for the finite radius of the cross-section of a vortex torus gives a lower estimate of the magnitude of the side lobes of the directivity pattern.

Now consider in short the acoustic scattering at vortex oscillations. It is well-known (Dhanak and De Bernardinis, 1981) that the oscillations may be excited at different modes in a vortex ring. The simplest of them (flexural waves) may be investigated in the approximation of an infinitely thin vortex sheet:

$$\Omega(y, r_1) = \Gamma\delta[y - y_0(t) - \varepsilon_y \sin\omega_0 t\cos m\psi]\delta(r_\perp - r_0 - \varepsilon_r\cos\omega_0 t\cos m\psi), \quad (8.48)$$

where $\varepsilon_y, \varepsilon_r \ll L$ and $m = 1, 2, 3, \ldots$ is the number of the mode.

Consider the acoustic scattering at the second mode $m = 2$. The analysis is restricted to the case when the vectors $\mathbf{U}$, $\mathbf{n}$, and $\mathbf{n}_0$ lie in the same plane. Then, the amplitude of scattering may be written as

$$f(\mathbf{n}, \mathbf{n}_0) = \frac{\Gamma\omega L}{2c_s^2} \frac{\cos\beta\cos(\beta/2)}{\sin(\beta/2)} \frac{\varepsilon_r}{L} \cos(\omega_0 t)\exp\left(\frac{i\omega q_y}{c_s} y_0(t)\right)$$

$$\times \left\{ \frac{1}{2} J_1(\xi) + J_3(\xi) - i\xi \left[J_4(\xi) - 2J_2(\xi) + J_0(\xi) \right] \right\}, \quad (8.49)$$

where $\xi = \omega L q_\perp / c_s$. Thus, the spectrum of the scattered signal must contain the combination frequencies $\omega + \Delta\omega \pm \omega_0$, where $\Delta\omega$ is the Dopler frequency shift emerging as a result of the translational vortex motion and ω_0 is the frequency of intrinsic oscillations of the vortex.

The directivity patterns of sound scattering by an oscillating vortex ring are shown in Fig. 8.6a and Fig. 8.6b for combination (dashed curve) and fundamental (solid curve) frequencies. These directivity patterns differ by the number and width of the lobes.

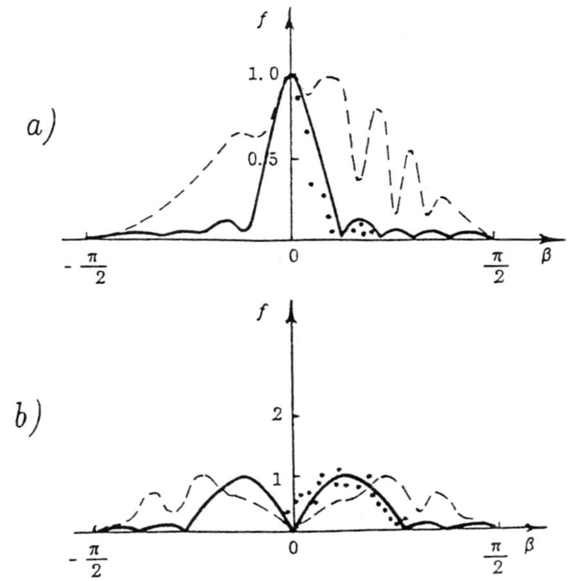

Figure 8.6: Normalized amplitude of acoustic scattering by an oscillating vortex ring. The dots stand for the values obtained in experiments.

Experimental investigation of sound scattering at a vortex ring was carried out using the same procedure as for the vortex wake (see Section 8.3.1). Vortex rings were produced by means of a section of a tapering cone closed by a shield with a round opening at one end and by a diaphragm of a dynamic loudspeaker at the other. Saw-tooth voltage was fed to the loudspeaker as a result of which the vortex rings were periodically escaping from the opening. The experiments were made on vortex rings at Reynolds numbers Re $= 2LU_0/\nu = 10^4$.

The velocity of vortex motion measured by the magnitude of the Dopler shift of the scattered sound differed by no more than 2 from that measured by means of a hot-wire anemometer. In Fig. 8.6 marked are the experimental magnitudes of the scattering amplitude $|f(\beta)|$ normalized to the maximal value. The absolute value $|f|$ may be calculated from (8.45) knowing the parameters a and L of the vortex ring. The circulation Γ may be determined by the data of acoustic sensing. It can also be calculated from data of contact measurements of the parameters a and L by means of the formula for the velocity of a thin toroidal vortex (Lamb, 1932):

$$U_0 = \frac{\Gamma}{4\pi L}\left(\log\frac{8L}{a} - \frac{1}{4}\right). \tag{8.50}$$

Like for the vortex wake, the accuracy of remote acoustic sensing of the absolute value of vortex intensity was about 16% at $\theta_0 = \pi/2$.

Results of experimental investigations indicate that the parameters of vortex rings may be determined by measurements of the scattered sound. Note that the sound with the wavelength on the order of ring diameter should be used in this case. At greater wavelengths, the amplitude of scattering becomes small and the scattering of the sound with a rather small wavelength is determined by small-scale turbulent pulsations rather than by the mean velocity field.

8.4 Internal gravity wave scattering by dipole vortices

The effects of scattering considered above are interesting for remote diagnostics of geophysical vortex flows by acoustic-gravity and internal waves (Ostrovsky, 1984). At the same time, the processes of scattering at vortices of the internal waves generated by different sources in real conditions may play a significant role in the dynamics of meso-scale phenomena in an atmosphere and in the ocean. This section is concerned with results of computations of the characteristics of IGW scattering by some typical vortex flows in a stratified medium.

The vorticity field described by a dipole vortex (the only isolated direction) is frequently taken as a model for vortices observed in the ocean and in the atmosphere. The Fourier transform of such a field

$$Q(\mathbf{k}) = VL^2(\mathbf{n}, \mathbf{k})W(kL) \tag{8.51}$$

is described by the unit vector $\mathbf{n} = (\cos\tilde{\varphi}, \sin\tilde{\varphi})$, characteristic velocity V and size L, and the dimensionless function $W(kL)$ of the scalar argument. The total vorticity for such flows is

$$\int Q\, d\mathbf{r} = Q_{\mathbf{k}=0} = 0$$

and the dipole moment is $\mathbf{P} = \mathbf{n}VL^2W(0)$.

Let us orient the x-axis along the wave vector of the incident wave. Then, the specific scattering cross-section for the dipole vortices has the form (Stepanyants and Fabrikant, 1991):

$$d\sigma = \frac{V^2L^4k^5}{8\pi\omega^2}\frac{[\cos(\varphi-\tilde{\varphi}) - \cos\tilde{\varphi}]^2}{\tan^2(\varphi/2)}\Phi^2(\varphi)W^2(2kL\sin\frac{\varphi}{2})\, d\varphi. \tag{8.52}$$

From this one can readily find the total scattering cross-section for long waves at $kL \ll 1$:

$$\sigma = \frac{L(kL)^3k^2V^2W^2(0)}{\pi\omega^2}\int_{-\pi/2}^{+\pi/2}\cos^2\varphi\,\sin^2(\varphi-\tilde{\varphi})\Phi^2(2\varphi)\, d\varphi. \tag{8.53}$$

Consider also the limiting case of short waves $kL \to \infty$. The total scattering cross-section is determined, primarily, by the scattering at small angles $\varphi \leq (kL){-}1$ and has the form

$$\sigma = \frac{L(kL)^2k^2V^2\Phi^2(0)}{\pi\omega^2}\sin^2\tilde{\varphi}\int_0^\infty W^2(z)\, dz. \tag{8.54}$$

Whereas for $\tilde{\varphi} = 0$, instead of this expression, at $kL \to \infty$ one has

$$\sigma = \frac{Lk^2V^2\Phi^2(0)}{4\pi\omega^2}\cos^2\tilde{\varphi}\int_0^\infty z^2W^2(z)\,dz. \tag{8.55}$$

Consider in more detail two broadly used models of vortex flows: the Lamb vortex (Lamb, 1932) and the Larichev–Reznik (LR) vortex (Larichev and Reznik, 1976). The vorticity for the Lamb vortex is written as

$$Q(r) = \frac{y}{r}\begin{cases} -\dfrac{2\kappa_0 V \mathrm{J}_1(\kappa_0 r/L)}{\mathrm{J}_1'(\kappa_0)}, & \text{for} \quad r \le L, \\ 0, & \text{for} \quad r > L. \end{cases} \tag{8.56}$$

Hereinafter, $\mathrm{J}_i(\xi)$ are referred to as the Bessel functions of the i-th kind and κ_0 as the first root of the equation $\mathrm{J}_1(\kappa_0) = 0$. The Fourier transform of this field may be represented by (8.51), where

$$W(kL) = -\frac{2i\mathrm{J}_1(kL)}{kL(1 - k^2L^2/\kappa_0^2)}. \tag{8.57}$$

Analogously, for the LR vortex one has

$$Q(r) = -\frac{LVy}{\mathrm{Ro}^2 r}\begin{cases} \dfrac{\mathrm{J}_1(\nu r/L)}{\mathrm{J}_1(\nu)}, & \text{for} \quad r \le L, \\ \dfrac{K_1(r/\mathrm{Ro})}{K_1(L/\mathrm{Ro})}, & \text{for} \quad r > L, \end{cases} \tag{8.58}$$

where Ro is the Rossby–Obukhov radius, $K_i(\xi)$ is the modified Bessel function of the i-th kind, and the parameter ν is found from the equation

$$\frac{\mathrm{J}_2(\nu)}{\nu\mathrm{J}_1(\nu)} + \frac{K_2(L/\mathrm{Ro})}{(L/\mathrm{Ro})K_1(L/\mathrm{Ro})} = 0. \tag{8.59}$$

It should be emphasized that the wave motions in a rotating fluid in the f-plane approximation, i. e., omitting the β-effect (Kamenkovich *et al.*, 1987) are considered here. The LR vortices exist in this case, again, at finite values of Ro. The limit transition from the LR vortex to the Lamb vortex occurs as $L/\mathrm{Ro} \to 0$. The vorticity and velocity field for the LR vortex at distances $r \ll \mathrm{Ro}$ are close to the corresponding fields for the Lamb vortex. However, at $r > \mathrm{Ro}$, the vorticity field (8.58), unlike the vorticity for the Lamb vortex (8.56), is not equal to zero, instead it decreases smoothly as $e^{-r/\mathrm{Ro}}/\sqrt{r}$. The velocity field of the LR vortex decreases by the same law, while the velocity field of the Lamb vortex decreases as r^{-2}.

The amplitude and scattering cross-section were calculated numerically for both models (Stepanyants and Fabrikant, 1991). The characteristics of wave scattering at these two vortices are in a good qualitative agreement. Acoustic and internal gravity waves were considered primarily for $\Omega_0/\omega \to 0$. Two limiting cases: horizontally propagating acoustic waves and low-frequency IGWs for which $\omega \ll N$ were considered in the first place. As an example, the normalized indicatrix of scattering $|f(\varphi)|^2/|f_{\max}|^2$ for acoustic waves ($c_s k_0/\omega = 1$, $N = 0$, $\Omega_0 = 0$) is depicted in Fig. 8.7 for different values of the parameter $\alpha \equiv k_0 L$ and of the slope $\tilde{\varphi}$ of the dipole axis relative to the direction of the incident wave. Note that the shape of the indicatrix of scattering remained almost unchanged with the variation of the parameter α from 0.1 to 1.

The total scattering cross-section depends on the angle of incident wave with respect to the axis of vortex dipole $\tilde{\varphi}$ and increases with increasing parameter α. In Fig. 8.8 normalized scattering cross-section $S = \sigma\omega^2/k^2V^2L$ of sound waves by the Lamb vortex is depicted as a function of $\tilde{\varphi}$ (left) for different values of α and as a function of parameter α (right) for different values of $\tilde{\varphi}$. The most complete picture of the $S(\tilde{\varphi}, \alpha)$ dependence is presented in Fig. 8.9. The magnitude of S remains limited at $\alpha \to \infty$, provided that $\tilde{\varphi} = 0$, but increase ($S \sim \alpha^2$) at $\tilde{\varphi} \neq 0$. It should be remembered, however, that the relative scattering cross-section σ/L is much lesser than unity within the applicability limits of the perturbation theory considered here.

For high-frequency IGWs at $\omega \sim N$ and for the inertia-gravity waves with frequencies $\omega \sim \Omega_0$, the total scattering cross-section σ is almost the same as for the IGWs with $\Omega_0 \ll \omega \ll N$.

The scattering characteristics of the inertia-gravity wave ($c = \infty$, $\Omega_0/\omega = 0.2$, $N/\omega = 100$) by the LR vortex are presented in Fig. 8.10 and Fig. 8.11. Figure 8.10 shows the normalized indicatrix of scattering as a function of the parameter L/Ro.

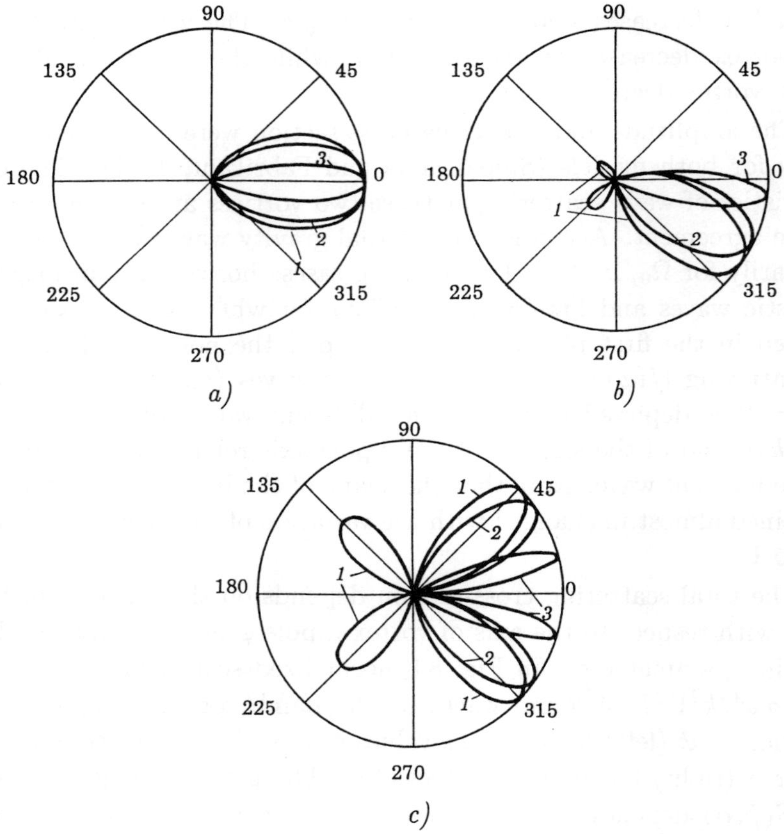

Figure 8.7: Normalized indicatrix of scattering of sound waves by the Lamb vortex at $\alpha = 1, 3, 10$ (curves 1,2,3): (a) $\tilde{\varphi} = 0$, (b) $\tilde{\varphi} = 5\pi/12$, (c) $\tilde{\varphi} = \pi/2$.

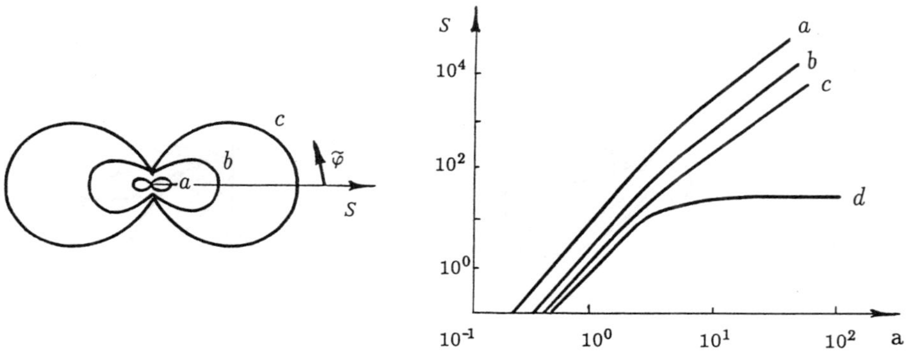

Figure 8.8: Total scattering cross-section S of sound waves by the Lamb vortex as a function of angle $\tilde{\varphi}$ (left) for different values of α: (a) $\alpha = 1$, (b) $\alpha = 1.5$, (c) $\alpha = 2$ and as a function of parameter α (right) for different values of $\tilde{\varphi}$: (a) $\tilde{\varphi} = 0$, (b) $\tilde{\varphi} = \pi/3$, (c) $\tilde{\varphi} = 5\pi/12$, (d) $\tilde{\varphi} = \pi/3$.

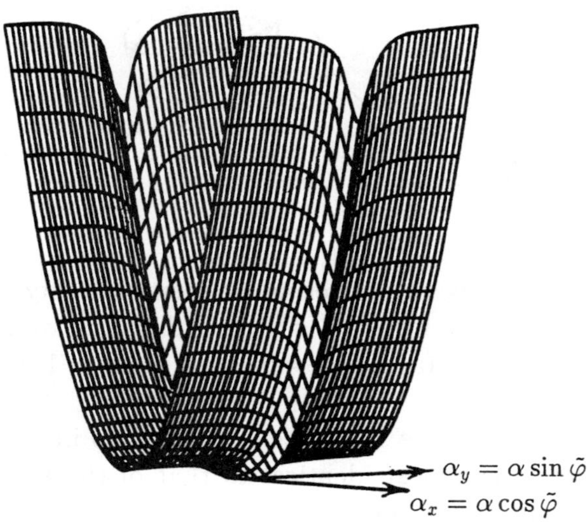

Figure 8.9: General form of the function $S(\alpha, \tilde{\varphi})$ in Cartesian coordinates: $\alpha_x = \alpha \cos \tilde{\varphi}$, $\alpha_y = \alpha \sin \tilde{\varphi}$. (The scale along α_y is compressed by a factor of ten as compared to the scale along α_x.)

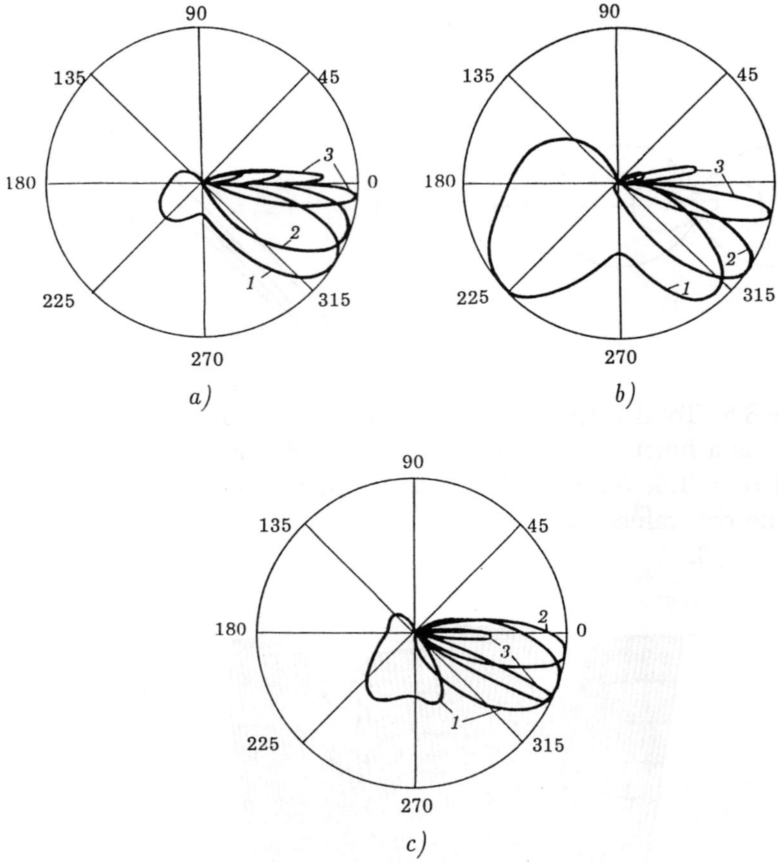

Figure 8.10: Indicatrix of scattering of inertia-gravity waves by the Larichev–Reznik vortex for $\tilde{\varphi} = \pi/3$ and different values of α [$\alpha = 1, 3, 10$ (curves 1, 2, 3)]: (a) $L/\text{Ro} = 0.1$, (b) $L/\text{Ro} = 1$, (c) $L/\text{Ro} = 10$.

At $L/\text{Ro} \to 0$ when the LR vortex degenerates into the Lamb vortex, the indicatrix is almost identical to that of the Lamb vortex (cf. Fig. 8.7). The only significant difference is the presence of a dip in the plot for the indicatrix at $\varphi \to 0$, which corresponds to a sharp decrease of forward scattering. This effect has a simple explanation. The scattering at small angles is determined by the deflection in the vortex field of the rays with a larger impact parameter. If the rays traverse the region

$r \leq$ Ro where the velocity field at $L/\text{Ro} \ll 1$ is nearly equal to that of the Lamb vortex, then the scattering is determined by the ray deflection in this region and depends slightly on the scattering by the Lamb vortex. At the same time, the rays making their path at a distance $r >$ Ro from the vortex axis where the velocity field decreases exponentially, are deflected only slightly, hence, the amplitude of scattering at small angles is much smaller for the LR vortex than for the Lamb vortex.

As Ro is increasing, the dip in the indicatrix of scattering vanishes the faster, the smaller α. At large α, the side lobes appear in the indicatrix with increasing Ro, and the directivity pattern of the scattered wave now has a three-lobe pattern.

Finally, Fig. 8.11 plots the normalized total scattering cross-section S of long IGWs ($N \gg \omega \gg \Omega_0$) versus L/Ro at the LR vortex for

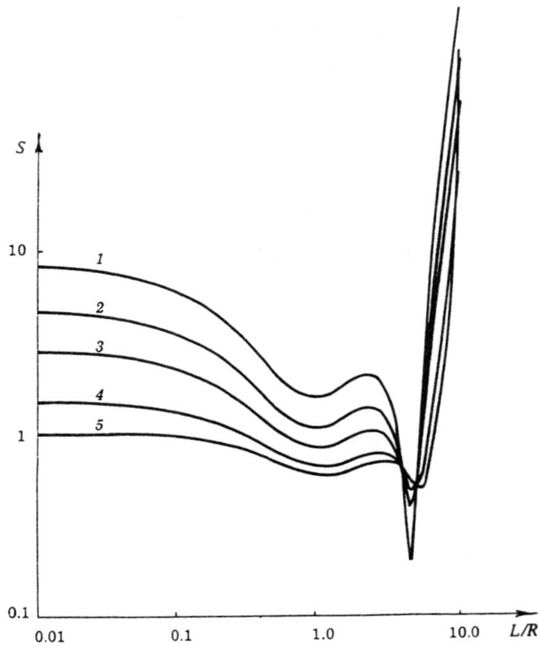

Figure 8.11: Total scattering cross-section S of the inertia-gravity waves by the Larichev–Reznik vortex.

$\alpha = 1$ and different values of $\tilde{\varphi}$: 1 – $\tilde{\varphi} = 0$; 2 – $\tilde{\varphi} = \pi/4$; 3 – $\tilde{\varphi} = \pi/3$; 4 – $\tilde{\varphi} = 5\pi/12$; 5 – $\tilde{\varphi} = \pi/2$.

As the parameter L/Ro is increased, S first decreases from the value corresponding to the Lamb vortex to some minimal value at $L/\mathrm{Ro} \sim 1$. As this parameter increases still further, S increases without restriction $(L/\mathrm{Ro})^4$. This is the result of the assumption that the vortex velocity is constant and nonzero. Another particular case is a possible transition to the Stern vortex (Stern, 1975) that is an interesting example of the vortex motion inside the cylindrical region outside which the fluid is at rest. This flow is formed from the LR vortex at $V \to 0$, $L/\mathrm{Ro} \to \infty$, and $V_{eff} \equiv VL^2/R_0^2 = \mathrm{const}$. The quantity $S_{eff} \equiv \sigma\omega^2/V_{eff}^2 k^2 L$ tends to a constant in this limit.

In conclusion let us estimate the magnitude of the cross-section of IGW scattering by mesoscale vortices in an atmosphere. Let us take the following typical parameters: $V \sim 15$ m/s, $L \sim 10$ km, $\lambda = 2\pi/k \sim 100$ km, $T = 2\pi/\omega \sim 10$ h, and $N \sim 10^{-2}$ s^{-1}. The scattering cross-section for the dipole-type vortices in this case is about $0.06L$. For the IGWs propagating against the background of mesoscale turbulence that consists of such vortices with random phases, the characteristic scale of damping of the mean wave field is $l_\nu \sim L^2/\sigma \sim 150$ km. Thus, allowance for the IGW scattering by barotropic vortex flows may be useful for the description of wave phenomena in an atmosphere.

BIBLIOGRAPHY

Abramovich, B. S., Mareev, E. A., Nemtsov, B. E. [1986] "Oscillatory instability of a moving oscillator radiating surface and internal waves", Izv. Akad Nauk SSSR, Mekh. Zhidkosti i Gaza., **1**, 168–171 (in Russian).

Acheson, D. J. [1976] "On over-reflection", J. Fluid Mech., **77**, 3, 433–472.

Ahmed, B. M., Eltayeb, L. A. [1988] "On the propagation, reflection, transmission and stability of atmospheric Rossby-gravity waves on a beta-plane in the presence of latitudinally shear flows", Phil. Trans. Roy. Soc. **A 298**, 1435, 45–85.

Alpar, M. A., Anderson, P. W., Pines, D., Shaham, J. [1981] "Giant glitches and pinned vorticity in the Vela and other pulsars", Astrophys. J., **249**, 1, pt. 2, 29–33.

Andronov, A. A., Chugunov, Yu. V., [1975] "Quasistationary electric fields of sources in a rarefied plasma", Usp. Fiz. Nauk, **116**, 1, 79 (in Russian).

Andronov, A. A., Fabrikant, A. L. [1979] "Landau damping, wind waves and whistles", in: Nonlinear Waves (ed. by A. V. Gaponov-Grekhov), Nauka, Moscow, pp. 68–104 (in Russian).

Andronov, A. A., Fabrikant, A. L. [1980] "Theory of aerodynamic self-excitation of sound: amplification of surface waves", Sov. Phys. Acoust., **26**, 5, 370–374 [Engl. transl. from Akust. Zh., **26**, 5, 655 (1980)].

Andronov, A. A., Fabrikant, A. L. [1980] "Theory of aerodynamic self-excitation of sound: resonator whistlers", Sov. Phys. Acoust., **26**, 6, 461–464 [Engl. transl. from Akust. Zh., **26**, 6, 817 (1980)].

Aranson, I. S., Gorshkov, K. A., Rabinovich, M. I. [1982] The Stochastization of Soliton Motion in a Periodic Field, Preprint N 51, Inst. Appl. Phys., USSR Acad. Sci., Gorky, USSR, 23 pp. (in Russian).

Badulin, S. I., Shrira, V. I., Tsimring, L. Sh. [1985] "The trapping and vertical focusing of internal waves in a pycnocline due to the horizontal inhomogeneities of density and currents", J. Fluid Mech., **158**, 199–218.

Barnett, N. P., Kenyon, K. E. [1975] "Recent advances in the study of wind waves", Repts. Progr. Phys., **38**, 6, 667–729.

Basovich, A. Ya. [1978] "Amplification of surface waves reflected from a vortex sheet", Izv. Atmos. Ocean. Phys., **14**, 637 [Engl. transl. from Izv. Akad. Nauk SSSR, Fiz. Atmos. i Okeana, **14**, 8, 901–904 (1978)].

Basovich, A. Ya., Tsimring, L. Sh. [1984] "Internal waves in a horizontally inhomogeneous flow", J. Fluid Mech., **142**, 233–249.

Batchelor, J. K. [1970] An Introduction to Fluid Dynamics, Cambridge University Press.

Benjamin, T. B. [1963] "The threefold classification of unstable disturbances in flexible surfaces bounding inviscid flows", J. Fluid Mech., **16**, pt. 3, 436–450.

Benjamin, T. B. [1986] "Note on added mass and drift", J. Fluid Mech., **169**, 251–256.

Betchov, R., Criminale, W. O. [1967] Stability of Parallel Flows, Academic Press, N. Y. – London.

Blokhintsev, D. I. [1981] Acoustics of Inhomogeneous Moving Medium, Nauka, Moscow, 260 pp. (in Russian).

Blúmen, W., Drazin, P. G., Billings, D. F. [1975] "Shear layer instability of an inviscid compressible fluid. Part 2", J. Fluid Mech., **71**, 2, 305–316.

Bogolyubov, N. N., Mitropol'sky, Yu. A. [1963] Asymptotic Methods in the Theory of Nonlinear Oscillations, Fizmatgiz, Moscow (in Russian).

Booker, J. R., Bretherton, F. P. [1967] "The critical layer for internal gravity waves in a shear flow", J. Fluid Mech., **27**, 3, 513–539.

Boyarintsev, V. I., Leont'ev, A. I., Sekerzh-Zen'kovich, S. Ya., Sysoev, V. I. [1982] "The propagation of vortex rings in density inhomogeneous fluid", Zh. Prikl. Mat. i Tekhn. Fiz., 2, 22–26 (in Russian).

Bretherton, F. P. [1969] "On the mean motion induced by internal gravity waves", J. Fluid Mech., **36**, 4, 785–803.

Briggs, R. J., Daugherty, J. D., Levy, R. H. [1970] "Role of Landau damping in cross-field electron beams and inviscid shear flow", Phys. Fluids, **13**, 2, 421–432.

Broadbent, E. G., Moore, D. W. [1979] "Acoustic destabilization of vortices", Phil. Trans. Roy. Soc. London, **A 290**, 1372, 353–371.

Brown, G. B. [1937] "The mechanism of the edge-tone production", Proc. Roy. Soc. London, **49**, 493.

Busse, F. H. [1968] "Shear flow instabilities in rotating systems", J. Fluid Mech., **33**, 3, 577–589.

Cairns, R. A. [1979] "The role of negative energy waves in some instabilities of parallel flows", J. Fluid Mech., **82**, 1, 1–14.

Case, K. M. [1961] "Hydrodynamic stability and the inviscid limit", J. Fluid Mech., **10**, 3, 420.

Case, K. M. [1962] "Hydrodynamic stability as a problem with initial data", in: Hydrodynamic Instability, Proc. Symp. Appl. Math., v. XIII, ed. by G. Birkhoff, R. Bellman, C. C. Lin, Amer. Math. Soc.

Chan, V. C. [1977] "Rayleigh's method as applied to the problem of wind-generated surface waves on deep water", Izv. Akad Nauk SSSR, Mekh. Zhidk. i Gaza, **6**, 154–157 (in Russian).

Chanaud, R. C. [1970] "Aerodynamic whistlers", Sci. Am., **222**, 1, 40–46.

Chen, C. N. [1975] "60 Jahre Forshung über die Karmanschen Wirbelstrassen. Ein Rückblick", Schweitzerische Bauzeitung, **21**, 44, 1079–1096.

Chukbar, K. V., Yan'kov, V. V. [1977] "Langmuir solitons in an inhomogeneous plasma", Fizika Plazmy, **3**, 6, 1398–1401 (in Russian).

Coltman, J. W. [1968] "Sounding mechanism of the flute and organ pipe", J. Acoust. Amer., **44**, 4, 983–992.

Coy, L., Fritts, D. C., Weinstock, J. J. [1986] "The Stokes drift due to vertically propagating internal gravity waves in a compressible atmosphere", J. Atmos. Sci., **43**, 22, 2636–2643.

Craik, A. D. D. [1985] Wave Interactions and Fluid Flows, Cambridge University Press.

Dahm, W. J. A., Scheil, C. M., Tryggvason, G. J. [1989] "Dynamics of vortex interaction with a density interface", J. Fluid Mech., **205**, 1–44.

Danilov, S. D. [1989] "On acoustic instability of flows with closed streamlines", Akust. Zh., **35**, 6, 1059–1065 (in Russian).

Derr, V. E., Little, C. G. [1970] "A comparison of remote sensing of the clear atmosphere by optical, radio, and acoustical radar techniques", Appl. Optics, **9**, 9, 1976–1992.

Dhanak, M. R., De Bernardinis, B. [1981] "The evolution of an elliptic vortex ring", J. Fluid Mech., **109**, 189–219.

Dikiĭ, L. A. [1960] Sov. Phys. Dokl., **5**, 1179 [Engl. transl. from Dokl. Akad. Nauk SSSR, **135**, 5, 1068–1071 (1960)].

Dolina, I. S. [1984] "Enhancement of oscillatory motion of bodies in a stratified liquid", Isv. Akad. Nauk SSSR, Mekh. Zhidk. i Gaza, 4, 87–93 (in Russian).

Drazin, P. G., Reid, W. H. [1981] Hydrodynamic Stability, Cambridge University Press, 525 pp.

Dysthe, K. B. [1979] "Note on a modification to the nonlinear Schrödinger equation for application to deep waves", Proc. Roy. Soc. London, **A 369**, 105–114.

Eckhart, C. [1948] "Vortices and streams caused by sound waves", Phys. Rev., **73**, 1, 68.

Elder, S. A. [1978] "Self-excited depth-mode resonance for a wall-mounted cavity in turbulent flow", J. Acoust. Soc. Amer., **64**, 3, 877–890.

Eltayeb, I. A., McKenzie, J. F. [1975] "Critical layer behavior and wave amplification on a gravity wave incident upon a shear layer", J. Fluid Mech., **72**, 4, 661–671.

Engelbrecht, Yu. K., Fridman, V. E., Pelinovsky, E. N. [1988] Nonlinear Evolution Equations, Longmann, London, 122 pp.

Erokhin, N. S., Moiseev, S. S. [1973] "Some problems of linear and nonlinear wave transformation in inhomogeneous media", Usp. Fiz. Nauk, **109**, 2, 225–258 (in Russian).

Erokhin, N. S., Moiseev, S. S. [1974] "Wave transformation in an inhomogeneous unstable plasma", Sov. Phys. JETP, **38**, 712 [Engl. transl. from Zh. Eksp. Teor. Fiz., **65**, 1431–1447 (1973)].

Erokhin, N. S., Sagdeev, R. Z. [1985] "To the theory of anomalous focusing of internal waves in a two-dimensional non-uniform fluid. Pt. 1: A stationary problem", Morskoi Gidrofiz. Zh., 2, 15–27 (in Russian).

Erokhin, N. S., Sagdeev, R. Z. [1985] "On the theory of anomalous focus of internal waves in horizontally inhomogeneous fluid. Pt. 2: Precise solution of two-dimensional problem with regard for viscosity and non-stationarity", Morskoi Gidrofiz. Zh., 4, 3–10 (in Russian).

Evans, D. V. [1975] "The transmission of deep-water wakes across a vortex sheet", J. Fluid Mech., **68**, pt. 3, 389–401.

Ezersky, A. B., Ostrovsky, L. A., Stepanyants, Yu. A. [1981] "Induced flows and their contribution to the energy of fluid wave motions", Izv. Atmos. Ocean. Phys., **17**, 890 [Engl. transl. from Izv. Akad. Nauk SSSR, Fiz. Atmos. i Okeana, **17**, 11, 1201–1208 (1981)].

Ezersky, A. B., Papko, V. V. [1986] "Laboratory investigation of large-scale potential flows induced by a packet of surface waves", Izv. Atmos. Ocean. Phys., **22**, 756 [Engl. transl. from Izv. Akad. Nauk SSSR, Fiz. Atmos. i Okeana, **22**, 9, 979–986 (1986)].

Ezersky, A. B., Rabinovich, M. I., Fabrikant, A. L. [1983] A Method for Measurement of the Parameters of Large-Scale Structures in Turbulent Flows, Author's Certificate 1056042 MKU GO1N 29/00, Patent N 43, Bull. of Inventions (in Russian).

Ezersky, A. B., Stepanyants, Yu. A., Tsimring, L. Sh. [1981] The Role of Negative Energy Waves on Unstable Water Surface Affected by a Water Flow, Preprint N 16, Inst. Appl. Phys. (in Russian).

Fabrikant, A. L. [1976] "Resonant interaction of sound with a parallel shear flow", Sov. Phys. Acoust, **22**, 1, 56–59 [Engl. transl. from Akoust. Zh., **22**, 1, 107 (1976)].

Fabrikant, A. L. [1980] "On nonlinear development of wind waves under a light wind", Izv. Acad. Sci. USSR, Atmos. Oceanic Phys., **16**, 9 [Engl. transl. from Izv. Akad. Nauk SSSR, Fiz. Atmos. i Okeana, **16**, 9, 985 (1980)].

Fabrikant, A. L. [1980] Generation of Self-Excited Oscillations in Non-equilibrium Systems Containing Flows, Doctoral Thesis, Gorky State University (in Russian).

Fabrikant, A. L. [1982] "Sound scattering by a vortex", Sov. Phys. Acoust., **28**, 5, 410–411 [Engl. transl. from Akust. Zh., **28**, 5, 694 (1982)].

Fabrikant, A. L. [1983] "Sound scattering by vortex flows", Sov. Phys. Acoust., **29**, 2 [Engl. transl. from Akust. Zh., **29**, 2, 262 (1983)].

Fabrikant, A. L. [1984] "Nonlinear dynamics of wave packets in a dissipative medium", Sov. Phys. JETP, **59**, 2, 274–278 [Engl. transl. from Zh. Eksp. Teor. Fiz., **86**, 2, 470–478 (1984)].

Fabrikant, A. L. [1987] "The reflection of Rossby waves from the surface of a tangential velocity discontinuity", Izv. Acad. Sci. USSR, Atmos. Oceanic Phys., **23**, 1, 79–81 [Engl. transl. from Izv. Akad. Nauk SSSR, Fiz. Atmos. i Okeana, **23**, 1, 106–109 (1987)].

Fabrikant, A. L. [1987] "Super-reflection of waves in hydrodynamic flows", Radiophysics and Quantum Electronics, **30**, 2, 213–224 [Engl. transl. from Izv. VUZov, Radiofizika, **30**, 2, 275–288 (1987)].

Fabrikant, A. L. [1988] Momentum of the Wave Packet in a Medium and Induced Mean Flows, Preprint N 183, Inst. Appl. Phys. (in Russian).

Fabrikant, A. L. [1990] "Resonant amplification of the internal gravity waves scattered at a cylindrical vortex", in: Oscillations and Waves in Liquids and Gases, Gorky Polytechnical Inst., Gorky, pp. 70–79 (in Russian).

Feix, M. R. [1964] "Impedance of RF grids and plasma condensers", Phys. Rev. Lett., **12**, 4, 316–317.

Feijer, J. A. [1963] "Hydromagnetic reflection and refraction at a fluid velocity discontinuity", Phys. Fluids, **6**, 4, 508–512.

Ferziger, J. H. [1974] "Low-frequency acoustic scattering from a trailing vortex", J. Acoust. Soc. Amer., **56**, 6, 1705–1707.

Fetter, A. L., Harvey, K. [1971] "Elastic filaments and vortex waves", Phys. Rev., **4A**, 6, 2306–2308.

Fjørtøft, R. [1950] "Application of integral theorems in deriving criteria of stability of laminar flow and for the baroclinic circular vortex", Geophys. Publ., **17**, 6, 1–52.

Francis, S. H. [1975] "Global propagation of atmospheric gravity waves: A review", J. Atm. Terr. Phys., **37**, 6–7, 1011–1058.

Fritts, D. C. [1980] "Simple stability limits for vertically propagating unstable modes in a tanh(z) velocity profile with a rigid lower boundary", J. Atmos. Sci., **37**, 7, 1642–1648.

Galeev, A. A., Sagdeev, R. Z. [1973] "Nonlinear plasma theory", in: Problems of Plasma Theory, Atomizdat, Moscow, v. 7, pp. 3–145 (in Russian).

Gantmacher, F. R. [1966] Lectures on Analytical Mechanics, Nauka, Moscow (in Russian).

Gaponov, V. I. [1960] Electronics, Nauka, Moscow (in Russian).

Gaponov-Grekhov, A. V., Dolina, I. S., Ostrovsky, L. A. [1983] "Anomalous Dopler effect and radiative instability of oscillators motion in hydrodynamics" Dokl. Akad. Nauk SSSR, **268**, 4, 827–831 (in Russian).

Gasenko, V. G., Ivansky, A. P. [1979] "Solitary waves in an active medium with dispersion and dissipation. Waves on a vertical film", in: Hydrodynamics and Heat Exchange in One- and Two-Phase Media, Inst. Thermal Phys., Novosibirsk, pp. 95–101 (in Russian).

Gavrilenko, V. G., Zelekson, L. A. [1977] "Sound amplification in an inhomogeneous flow", Akust. Zh., **23**, 6, 867–872 (in Russian).

Gavrilenko, V. G., Zelekson, L. A. [1977] "Resonant interaction of electromagnetic waves and an inhomogeneous flow of magnetoactive plasma", Radiophys. Quantum Electron., **20** [Engl. transl. from Izv. VUZov, Radiofizika, **20**, 7, 982–986 (1977).

Gavrilenko, V. G., Zelekson, L. A., Basovich, A. Ya. [1975] "On the instability of a supersonic jet of an inviscid medium", in: Proc. Symp. on the Physics of Acousto-Hydrodynamic Phenomena (Sukhumi, 1985), Nauka, Moscow, pp. 250–253 (in Russian).

Gavrilov, N. M. [1987] "Nonlinear generation of wave motions by quasi-geostrophic and vortex motions in an atmosphere", Meteorolog. Issledovan., 3, 49–71 (in Russian).

Gertzenshtein, S. Ya. [1969] "On the convergence of Rayleigh's method", Dokl. Akad. Nauk SSSR, **187**, 5, 1012–1015 (in Russian).

Gerwin, R. A. [1968] "Stability of a vortex sheet", Rev. Mod. Phys., **40**, 3, 652–658.

Gestrin, S. G., Kontorovich, V. M. [1986] "Wind instability and spiral perturbations of relativistic magnetized jets", Zh. Eksp. Teor. Fiz., **91**, 3, 779–791 (in Russian).

Gill, A. E. [1982] Atmosphere–Ocean Dynamics, Academic Press, N. Y. – London – Paris – San Diego.

Ginzburg, V. L. [1967] Electromagnetic Wave Propagation in a Plasma, Nauka, Moscow (in Russian).

Ginzburg, V. L. [1975] Theoretical Physics and Astrophysics, Nauka, Moscow (in Russian).

Ginzburg, V. L., Ugarov, V. A. [1976] "Some notes on forces and electromagnetic tensor in macroscopic electrodynamics", Sov. Phys. Usp., **19**, 94 [Engl. transl. from Uspekhi. Fiz. Nauk, **118**, 1, 175–188 (1976)].

Gnevyshev, V. G., Shrira, V. I. [1989a] "The transformation of monochromatic Rossby waves in a critical layer of a horizontal flow", Izv. Akad. Nauk, Fiz. Atmos. i Okeana, **25**, 8, 852–862 (in Russian).

Gnevyshev, V. G., Shrira, V. I. [1989b] "The dynamics of Rossby wave trains in the neighbourhood of a zonal critical layer with viscosity taken into acoount", Izv. Akad. Nauk, Fiz. Atmos. i Okeana, **25**, 10, 1064–1074 (in Russian).

Gol'dshtik, M. A., Shtern, V. N. [1977] Hydrodynamic Instability and Turbulence, Nauka, Novosibirsk, 366 pp. (in Russian).

Goldstone, L. O., Oliner, A. A. [1959] "Leaky-wave antennas. Pt.1. Rectangular waveguides", IRE Trans., **AP 7**, 307.

Golemshtok, G. M., Fabrikant, A. L. [1980] "Scattering and amplification of sound waves by a cylindrical vortex", Sov. Phys. Acoust., **26**, 3, 209–213 [Engl. transl. from Akust. Zh., **26**, 3, 383 (1980)].

Gollub, J. P. and Swinney, H. L. eds. [1981] Hydrodynamic Instabilities and the Transition to Turbulence, Springer–Verlag, Berlin – Heidelberg – New York.

Golubyatnikov, G. Yu., Egorov, S. V., Kostrov, A. V., Mareev, E. A. [1989] "Quasistatic wave trapping into a heat waveguide formed by the near field of magnetic antenna in a magnetized plasma", Zh. Eksp. Teor. Fiz., **96**, 6, 2009–2017 (in Russian).

Golyamin, I. P., ed. [1979] Ultrasound: Small Encyclopaedia, Nauka, Moscow, 401 pp. (in Russian).

Gorev, V. V., Kingsep, A. S., Rudakov, L. I. [1976] "Long Langmuir plasma turbulence", Izv. VUZov, Radiofizika, **19**, 5–6, 691 (in Russian).

Gossard, E., Hooke, W. H. [1975] Waves in the Atmosphere, Elsevier Scientific Publ. Co., Amsterdam – Oxford – New York.

Gostintsev, Yu. A., Shatskikh, Yu. V., Ivanov, E. A. [1986] "Generation of longwave acoustic perturbations in an atmosphere by a turbulent thermocline", Dokl. Akad. Nauk SSSR, **287**, 3, 583–585 (in Russian).

Graham, B. B., Graham, E. B. [1969] "Effect of shear layer on plane waves of sound in a fluid", J. Acoust. Soc. Amer., **46**, 1–2, 169–175.

Gromov, P. R., Ezersky, A. B., Kiyashko, S. V., Fabrikant, A. L. [1982] Sound Scattering by a Toroidal Vortex. Theory and Experiment, Preprint N 59, Inst. Appl. Phys., Gorky (in Russian).

Gromov, P. R., Ezersky, A. B., Fabrikant, A. L. [1982] "Sound scattering at a vortex wake behind a cylinder", Sov. Phys. Acoust., **28**, 6, 452–455 [Engl. transl. from Akust. Zh., **28**, 6, 763 (1982)].

Gryanik, V. M. [1983] "Sound radiation by linear vortex filaments", Izv. Akad. Nauk SSSR, Fiz. Atmos. i Okeana, **19**, 2, 203–206 (in Russian).

Gyan Dyn Hua [1961] "To the theory of excitation of surface acoustic waves", Akust. Zh., **7**, 2, 181–184 (in Russian).

Hall, M. G. [1972] "Vortex breakdown", Annual Rev. Fluid Mech., **4**, 195–218.

Hasegawa, A., Mima, K. [1978] "Pseudo-three dimensional turbulence in magnetized nonuniform plasma", Phys. Fluids, **21**, 87–92.

Hasselmann, K. [1976] "Weak-interaction theory of ocean waves", in: Basic Developments in Fluid Dynamics, Academic Press, New York – London, v.2, pp. 117–182.

Høiland, E. [1953] "On two-dimensional perturbations of linear flow", Geofys. Publikasjoner, Norske Videnskaps–Acad., Oslo, **18**, 9, 1–12.

Holton, J. R. [1975] The Dynamic Meteorology of the Stratosphere and Mesosphere, Meteorological Monographs, V. 15, N 37, Amer. Meteorol. Soc.

Howard, L. N., Gupta, A. S. [1962] "On the hydrodynamic and hydromagnetic stability of swirling flows", J. Fluid Mech., **14**, 3, 463–476.

Howe, M. S. [1976] "The influence of vortex shedding on the generation of sound by convected turbulence", J. Fluid Mech., **76**, 4, 711–740.

Howe, M. S. [1983] "On the scattering of sound by a vortex ring", J. Sound and Vibr., **87**, 4, 567–571.

Ignat'ev, R. N., Sonin, E. B. [1981] "On a vortex finite-size lattice in a rotating superfluid", Zh. Eksp. Teor. Fiz., **81**, 6, 2059–2060 (in Russian).

Ince, E. L. [1956] Ordinary Differential Equations, Dover, N. Y.

Ivanovsky, A. I. [1959] Theoretical and Experimental Investigation of Sound Induced Flows, Gidrometeoizdat, Moscow (in Russian).

Jackson, J. D. [1962] Classical Electrodynamics, John Wiley and Sons Inc., New York – London.

Janke, E., Emde, F., Lösch, F. [1960] Tafeln Höherer Funktionen, B. G. Teubner Verlagschaft, Stuttgart.

Jones, W. L. [1968] "Reflection and stability of waves in stably stratified fluids with shear flows: a numerical study", J. Fluid Mech., **34**, 3, 609–624.

Jones, D. C., Morgan, J. D. [1972] "Instability of the vortex sheet under the action of sound", Proc. Camb. Phil. Soc., **72**, 3, 465–491.

Kadomtsev, B. B. [1988] Collective Phenomena in Plasma, Nauka, Moscow, 303 pp. (in Russian).

Kadomtsev, B. B., Mikhaĭlovskiĭ, A. B., Timofeev, A. V. [1964] "Negative energy waves in dispersive media", Zh. Eksp. Teor. Fiz., **47**, 6, 2266–2275 (in Russian).

Kambe, T., Minota, T., Ikushima, Y. J. [1985] "Acoustic wave emitted by a vortex ring passing near the edge of a half-plane", J. Fluid Mech., **155**, 77–103.

Kambe, T, Oo, U. M. [1981] "Scattering of sound by vortex rings", J. Phys. Soc. Jap., **50**, 10, 3507–3516.

Kamenkovich, V. M., Koshlyakov, M. N., Monin, A. S. [1987] Synoptic Vortices in the Ocean, Gidrometeoizdat, Leningrad, 511 pp. (in Russian).

Kapitsa, P. L., Kapitsa, S. P. [1949] "A wave flow of thin layers of viscous liquid. Pt. II. A flow contacting a gas flow and heat transfer", Zh. Eksp. Teor. Fiz., **19**, 2, 105–120 (in Russian).

Karpman, V. I. [1975] Nonlinear Waves in Dispersive Media, Pergamon Press, Oxford.

Karpman, V. I. [1979] "Soliton evolution in the pesence of a perturbation", Phys. Scripta, **20**, 462–478.

Kashkaryan, P. R., Frumkin, A. L. [1975] "Two-terminal networks with modulated parameters", Izv. VUZov, Radioelektronika, **18**, 8, 124 (in Russian).

Kashkaryan, P. R., Frumkin, A. L. [1976] "A parametric element containing a two-terminal network with modulated capacitance", Izv. VUZov, Radioelektronika, **19**, 8, 101–102 (in Russian).

Kikina, N. G. [1967] "On the effect of viscosity on vortex sheet instability in an incompressible medium", Sov. Phys. Acoust., **13**, 284 [Engl. transl. from Akust. Zh. **13**, 2, 213–217 (1967)].

Kikina, N. G., Sannikov, D. G. [1969] "Sound wave reflection at a moving plane-parallel flow", Akust. Zh., **15**, 4, 543–546 (in Russian).

Kiyashko, S. V., Papko, S. V., Rabinovich, M. I. [1975] "Model experiments on the interaction of Langmuir and ion-acoustic waves", Fizika Plazmy, **1**, 6, 1013 (in Russian).

Klyatskin, V. I. [1966] "Sound radiation by an ensemble of vortices", Izv. Akad. Nauk SSSR, Mekh. Zhidk. i Gaza, 6, 87–92 (in Russian).

Kochin, N. E., Kibel', I. A., Rose, N. V. [1963] Theoretical Hydrodynamics, Fizmatgiz, Moscow , V. 1,2 (in Russian).

Kolykhalov, P. I. [1984] "Vortex sheet instability in the gas moving near a rigid wall", Izv. Akad. Nauk SSSR, Fiz. Atmos. i Okeana, 3, 145 (in Russian).

Komatsu, T., Imasaka, M., Tsubaki, T. [1978] "Statistical properties of internal waves at the interface of two-layered flows", Techn. Report Kuishi Univ., V. 52, N 1, 25–32.

Konstantinov, B. P. [1974] Hydrodynamic Formation of Sound and Acoustic Propagation in a Bounded Medium, Nauka, Leningrad (in Russian).

Kop'ev, V. F., Leont'ev, E. A. [1983] "On acoustic instability of axial vortex", Sov. Phys. Acoust., **39**, 111 [Engl. transl. from Akust. Zh., **29**, 2, 192–198 (1983)].

Korpel, A., Banerjee, P. P. [1984] "A heuristic guide to nonlinear dispersive wave equations and soliton-type solutions", Proc. IEEE, **72**, 9, 1109–1130.

Korotaev, G. K. [1988] Theoretical Modelling of Synoptic Variability of the Ocean, Naukova Dumka, Kiev, 160 pp. (in Russian).

Krasnenko, N. P. [1986] Acoustic Sounding of the Atmosphere, Nauka, Novosibirsk, 164 pp. (in Russian).

Kuramoto, Y., Tsuzuki, T. [1976] "Persistent propagation of concentration waves in dissipative media far from thermal equilibrium", Progr. Theor. Phys., **55**, 2, 356–369.

Kurin, V. V., Nemtsov, B. E., Eidman, V. Ya. [1985] "Precursor and lateral waves at pulse reflection from the interface between two media", Sov. Phys. Usp., **28**, 827 [Engl. transl. from Uspekhi Fiz. Nauk, **147**, 1, 157–180 (1985)].

Lalas, D. P., Einaudi, F. J. [1976] "On the characteristics of gravity waves generated by atmospheric shear layers", J. Atmos. Sci., **33**, 7, 1248–1259.

Lamb, H. [1932] Hydrodynamics, Cambridge University Press.

Landau, L. D. [1969] Collected Papers, V. 2, Nauka, Moscow (in Russian).

Landau, L. D., Lifshits, E. M. [1973] Theory of the Field, Nauka, Moscow (in Russian).

Landau, L. D., Lifshits, E. M. [1973] Mechanics, Nauka, Moscow (in Russian).

Landau, L. D., Lifshits, E. M. [1986] Fluid Mechanics, Pergamon Press, London.

Larichev, V. D., Reznik, G. M. [1976] "Two-dimensional Rossby soliton: an exact solution", Polymermode News, **19**, 3, 6.

Larson, T. R., Wright, J. W. [1975] "Wind-generated gravity-capillary waves: laboratory measurements of temporal growth rates using microwave backscatter ", J. Fluid Mech., **70**, 3, 417–436.

Lavrent'ev, M. A., Shabat, B. V. [1977] Problems of Hydrodynamics and Their Mathematical Models, Nauka, Moscow, 477 pp. (in Russian).

LeBlond, P. H., Mysak, L. A. [1978] Waves in the Ocean, Elsevier, Amsterdam – Oxford – New York.

Lessen, M., Deshpande, N. V., Hadji-Ohenes, B. [1973] "Stability of a potential vortex with a non-rotating and rigid-body rotating top-hat jet core", J. Fluid Mech., **60**, 3, 459–466.

Lifshits, A. E. [1989] "On the echo phenomenon in hydrodynamics", Dokl. Akad. Nauk SSSR, **308**, 2, 340–344 (in Russian).

Lin, C. C. [1955] The Theory of Hydrodynamic Stability, Cambridge University Press.

Lin, S. P. [1974] "Finite amplitude side-band stability of a viscous film", J. Fluid Mech., **63**, 3, 417–429.

Lindzen, R. S., Barker, J. W. [1985] "Instability and wave over-reflection in stably stratified shear flow", J. Fluid Mech., **151**, 189–218.

Lindzen, R. S., Tung, K. K. [1978] "Wave over-reflection and shear instability", J. Atmos. Sci., **35**, 1626–1632 (in Russian).

Little, K. [1969] "Acoustic sounding of the lower atmosphere", Proc. IEEE, **57**, 4, 571–579.

Lupanov, G. A. [1975] "Behaviour of electromagnetic waves at a 'superluminiscent' velocity discontinuity in a medium", Radiophys. Quantum Electron., **18**, 1263 [Engl. transl. from Izv. VUZov, Radiofizika, **18**, 1711–1713 (1975)].

Lyamshev, L. M., Skvortsov, A. T. [1988] "Sound radiation by localized vortices in a weakly compressible medum", Akust. Zh., **34**, 5, 769–790.

McAllister, L. G., Pollard, J. R., Mahoney, A. R., Shaw, P. J. R. [1969] "Acoustic sensing: A new method for the investigation of the structure of atmosphere", Proc. IEEE, **57**, 4, 231–239.

McIntyre, M. E. [1981] "On the 'wave momentum' myth", J. Fluid Mech., **106**, 331–347.

McKenzie, J. F. [1972] "Reflection and amplification of acoustic-gravity waves at a density and velocity discontinuity", J. Geophys. Res., **77**, 16, 2915–2926.

Mercer, D. M. A. [1951] "Sound of organ pipes", JASA, **23**, 45–54.

Mercer, D. M. A. [1953] "The physics of the organ flue pipe", Amer. J. Phys., **21**, 5, 376–286.

Meyer, E. [1963] "Neuere Analogien zwischen akustischen und elektromagnetischen Schingungen und Wellenfeldern", in: IV Intern. Congr. Acoustics Congress Reports (Copenhagen 1962), V. 2, 139–156.

Michalke, A., Timme, A. [1967] "On the inviscid instability of certain two-dimensional vortex-type flows", J. Fluid Mech., **29**, 4, 647–666.

Mikhaĭlovskiĭ, A. B. [1974] Theory of Plasma Instabilities, Consultants Bureau, N. Y.

Miles, J. W. [1954] "Dispersive reflection at the interface between ideal and viscous media", JASA, **26**, 6, 1015–1018.

Miles, J. W. [1957] "On the reflection of sound at an interface of relative motion", JASA, **29**, 2, 226–228.

Miles, J. W. [1957] "On the generation of surface waves by shear flows", J. Fluid Mech., **3**, 2, 185–204. (See also Corrigenda [1959], 6, 582).

Miles, J. W. [1958] "On the disturbed motion of a plane vortex sheet", J. Fluid Mech., **4**, 538–552.

Miles, J. W. [1959] "On the generation of surface waves by shear flows", J. Fluid Mech., **6**, 4, 568–582.

Miles, J. W. [1967] "Surface-wave scattering matrix for a shelf", J. Fluid Mech., **28**, 4, 755–767.

Mironov, M. A. [1975] "The influence of a bulk-velocity harmonic source on a flow with a plane vortex sheet", Sov. Phys. Acoust., **21**, 48 [Engl. transl. from Akust. Zhurn., **21**, 1, 79–85 (1975)].

Mohring, W. [1978] "On vortex sound at low Mach number", J. Fluid Mech., **85**, 4, 685–691.

Moiseev, S. S., Pungin, V. G., Sagdeev, R. Z., Suyazov, N. V., Tur, A. V., Yanovsky, V. V. [1983] "Secondary and correlation instabilities in distributed systems", in: Nonlinear Waves, Self-Organization, Nauka, Moscow, pp. 85–95 (in Russian).

Nakoryakov, V. E., Shreiber, I. P. [1973] "Waves on the surface of a thin layer of viscous fluid", Prikl. Matem. i Tekhn. Fiz., **2**, 2, 109–113 (in Russian).

Nemtsov, B. E. [1986] Configuration Instability at Reflection of a Cylindrical Pulse from a Vortex Sheet, Preprint N 211, Res. Radiophysical Inst., Gorky.

Nemtsov, B. E., Eidman, V. Ya. [1982] "The effect of spatial echo in liquid with gas bubbles", Akust. Zh., **28**, 5, 669–672.

Nemtsov, B. E., Tsimring, L. Sh. [1986] Evolution of Initial Perturbations at the Interface of Two Nonmixing Flows, Preprint N 156, Inst. Appl. Phys., Gorky.

Nesterov, S. V., Sekerge-Zen'kovich, S. Ya. [1976] "Development of Kelvin instability in a nonstationary flow", Izv. Akad. Nauk SSSR, Fiz. Atmos. i Okeana, **12**, 10 (in Russian).

Nesterov, S. V., Sekerge-Zen'kovich, S. Ya. [1979] "Instability of the interface between flows at fluctuations of their relative velocity",

Izv. Akad. Nauk SSSR, Fiz. Atmos. i Okeana, **15**, 9, 947–952 (in Russian).

Nezlin, M. V. [1976] "Negative energy waves and anomalous Dopler effect", Sov. Phys. Usp., **19**, 946 [Engl. transl. from Usp. Fiz. Nauk, **120**, 3, 481–495 (1976)].

Nezlin, M. V., Snezhkin, E. N. [1993] Rossby Vortices and Spiral Structures, Springer-Verlag, Berlin – Heidelberg.

Novikov, A. A. [1977] "Nonlinear capillary waves on the surface of a jet of viscous fluid", Izv. Akad. Nauk SSSR, Mekh. Zhidkosti i Gaza, 2, 179–182 (in Russian).

Olbers, D. J. [1981] "The propagation of internal waves in a geostrophic current", J. Phys. Oceanogr., **11**, 9, 1224–1233.

O'Shea, S. [1975] "Sound scattering by a potential vortex", J. Sound Vibr., **43**, 1, 109–116.

Ostrovsky, L. A. [1968] "On second-order quantities in a progressive sound wave", Akust. Zh., **14**, 1, 82.

Ostrovsky, L. A. [1974] "Approximate methods in the theory of nonlinear waves", Izv. VUZov, Radiofizika, **17**, 4, 454–476 (in Russian).

Ostrovsky, L. A. [1984] "Surface and internal waves as a means for diagnostics of the ocean", in: Methods of Hydrophysical Research, Inst. Appl. Phys., Gorky, pp. 185–203 (in Russian).

Ostrovsky, L. A., Papilova, I. A. [1974] "On nonlinear acoustic wind", Akust. Zh., **20**, 1, 79–86 (in Russian).

Ostrovsky, L. A., Rybak, S. A., Tsimring, L. Sh. [1986] "Negative energy waves in hydrodynamics", Sov. Phys. Usp., **29**, 1040 [Engl. transl. from Uspekhi Fiz. Nauk, **150**, 3, 417–437 (1986)].

Ostrovsky, L. A., Soustova, I. A. [1979] "The upper mixed layer of the ocean as an energy sink of internal waves", Okeanologiya, **19**, 973–981 (in Russian).

Ostrovsky, L. A., Stepanyants, Yu. A., Tsimring, L. Sh. [1983] "The interaction of internal waves and flows with turbulence in the ocean", in: Nonlinear Waves: Self-Organization, ed. by A. V. Gaponov-Grekhov and M. I. Rabinovich, Nauka, Moscow (in Russian).

Ostrovsky, L. A., Stepanyants, Yu. A., Tsimring, L. Sh. [1984] "Internal solitons in the ocean: amplification, absorption and collective behavior", in: Nonlinear and Turbulent Processes in Physics, Harwood Academic Publishers, Gordon and Breach, New York, pp. 933–955.

Ott, E., Mannheimer, W. M., Book, D. L., Boris, J. P. [1973] "Model equations for mode coupling saturation in unstable plasmas", Phys. Fluids, **16**, 6, 855–862.

Pains, D. [1980] "Pulsars and compact X-ray sources. Cosmic laboratories for the study of neutron stars and hadron matter" Uspekhi Fiz. Nauk, **131**, 3, 479–494 (in Russian).

Pedlosky, J. [1982] Geophysical Fluid Dynamics, Springer-Verlag, New York – Heidelberg – Berlin.

Peierls, S. R. [1975] "The momentum of light in the refracting medium", Proc. Roy. Soc. Lond., **347**, 1651,475–491.

Pelinovsky, E. N. [1971] "On the absorption of nonlinear waves in nonlinear media", Prikl. Matem. i Tekhn. Fiz., 2, 68 (in Russian).

Pelinovsky, E. N. [1979] "Wave propagation in a statically nonuniform ocean", in: Nonlinear Waves, ed. by A. V. Gaponov-Grekhov, Nauka, Moscow, pp. 331–356 (in Russian).

Pelinovsky, E. N., Friedman, V. E. [1974] "Explosive instability of nonlinear waves in media with negative viscosity", Prikl. Mat. i Mekh., **38**, 6, 991 (in Russian).

Petelin, M. I. [1985] "Space charge effects at the excitation of a resonant cavity by a stationary flow of excited classical oscillators", Izv. Akad. Nauk SSSR, Fiz. Atmos. i Okeana, **28**, 3, 354–360 (in Russian).

Petrov, V. V. [1979] "To the dynamics of nonlinear surface waves in a stratified ocean", Izv. Akad. Nauk SSSR, Fizika Atmos. i Okeana, **15**, 7, 740 (in Russian).

Petviashvili, V. I. [1977] "Instability of a flow of ideal fluid with free surface", Dokl. Akad. Nauk SSSR, **234**, 4, 787–789 (in Russian).

Phillips, O. M. [1977] The Dynamics of the Upper Ocean, Cambridge University Press.

Pierce, J. [1974] Almost All About Waves, The MIT Press: Cambridge, Massachusetts and London, England.

Pitaevsky, L. P. [1958] "Calculation of the phonon part of the force of mutual friction in a superfluid helium", Zh. Eksp. Teor. Fiz., **35**, 5, 1271–1275 (in Russian).

Polyachenko, V. L., Fridman, A. M. [1976] The Equilibrium and Stability of Gravitating Systems, Nauka, Moscow, 447 pp. (in Russian).

Ponstein, J. [1959] "Instability of rotating cylindrical jets", Appl. Sci. Res., **A8**, 6, 425–456.

Rabinovich, M. I., Fabrikant, A. L. [1976] "Nonlinear waves in nonequilibrium media", Radiophys. and Quantum Electron., **19**, 5, 508–543 [Engl. transl. from Izv. VUZov, Radiofizika, **19**, 5–6, 721–766 (1976)].

Rabinovich, M. I., Reutov, V. P. [1973] "Interaction of parametrically coupled waves in nonequilibrium media", Izv. VUZov, Radiofizika, **16**, 8, 815–839 (in Russian).

Rabinovich, M. I., Trubetskov, D. I. [1989] Oscillations and Waves in Linear and Nonlinear Systems, Kluwer Academic Publishers, Dorderecht – Boston – London.

Rayleigh (J. W. Strutt) [1984] The Theory of Sound, V. III, Dover Press, N. Y., 1945.

Reutov, V. P. [1982] "Nonstationary critical layer and nonlinear stage of instability in a plane Poiseuille flow", J. Appl. Mech. Tech. Phys., **23**, 4 [Engl. transl. from Zh. Prikl. Mekh. Tekhn. Fiz., 4, 43–54 (1982)].

Ribner, H. S. [1957] "Reflection, transmission and amplification of sound by a moving medium", JASA, **29**, 435–441.

Ribner, H. S. [1985] "Cylindrical sound wave generated by shock-vortex interaction", AIAA Journ., **23**, 11, 1708–1715.

Rotunno, R. J. [1978] "A note on the stability of a cylindrical vortex sheet", J. Fluid Mech., **87**, 4, 761–771.

Rudenko, O. V., Soluyan, S. I. [1975] Theoretical Foundations of Nonlinear Acoustics, Nauka, Moscow, 287 pp. (in Russian).

Rudnitsky, A. G. [1989] "Remote acoustic determination of some parameters of turbulent flows", Akust. Zh., **35**, 1, 104–108 (in Russian).

Rudnitsky, A. G. [1990] "Sound transmission through the vortex wake behind a cylinder", Akust. Zh., **36**, 1, 122–129 (in Russian).

Russel, R. C., Onsorio, J. D. C. [1957] "An experimental investigation of drift profiles in a closed channel", in: Proc. 6-th Conf. Coastal Eng., Miami, 1957, pp. 171–193.

Russel, F. B., Uthe, E. F., Ludvig, F. L., Show, N. A. [1974] "A comparison of atmospheric structure as observed with monostatic acoustic sounder and lidar techniques", J. Geophys. Res., **79**, 36, 5555–5566.

Ryutov, D. D. [1975] "Analog of Landau damping in the problem of acoustic wave propagation in a fluid with gas bubbles", JETP Lett., **22**, 215 [Engl. transl. from Pis'ma v Zh. Eksp. Teor. Fiz., **22**, 9, 446 (1975)].

Shankar, P. H. [1981] "On the evolution of disturbances at an inviscid interface", J. Fluid Mech., **108**, 159–170.

Shchepetnov, R. V., Troitskaya, V. A., Dovbnya, B. V. [1986] "Electromagnetic radiation with central frequency of 2 Hz during powerful cyclone on June 9, 1984", Dokl. Akad. Nauk SSSR, **290**, 3, 582–585 (in Russian).

Shen, M. S., Shin, S. M. [1974] "Asymptotic theory of nonlinear waves on the surface of viscous fluid in inclined channel of arbitrary cross-section", Phys. Fluids., **17**, 2, 280–286.

Shirokova, N. L. [1970] "Coagulation of aerosoles", in: Physical Foundations of Ultrasound Technology, Nauka, Moscow (in Russian).

Shubert, L. K. [1972] "Numerical study of sound refraction by a jet flow", JASA, **51**, 1, 439–443.

Snow, J. T. [1978] "On inertial instability as related to the multiple-vortex phenomenon", J. Atmos. Sci., **35**, 9, 1660–1677.

Sonin, E. B. [1975] "Frictional force of normal component on vortices in a rotating superfluid helium", Zh. Eksp. Teor. Fiz., **69**, 3, 921–935 (in Russian).

Sretensky, L. N. [1977] Theory of Wave Motions of a Liquid, Nauka, Moscow, 915 pp. (in Russian).

Starobinsky, A. A. [1973] "Wave ampification at relative reflection from a rotating black hole", Zh. Eksp. Teor. Fiz., **64**, 48 (in Russian).

Steinmets, G. G., Singh, J. J. [1972] "Reflection and transmission of acoustical waves from a layer with space dependent velocity", JASA, **51**, 1, 218–222.

Stepanyants, Yu. A. [1983] Dynamics of Solitary Surface and Internal Waves in the Ocean, Ph. D. Thesis, Gorky, (in Russian).

Stepanyants, Yu. A., Fabrikant, A. L. [1989] "Propagation of waves in hydrodynamic shear flows", Sov. Phys. Usp., **32**, 9, 783–805 [Engl. transl. from Usp. Fiz. Nauk, **159**, 1, 83–123 (1989)].

Stepanyants, Yu. A., Fabrikant, A. L. [1991] "Sound and internal gravity waves scattering at concentrated vortices in a rotating stratified fluid", Izv. Atmos. Oceanic Phys., **27**, 2 [Engl. transl. from Izv. Akad Nauk SSSR, Fiz. Atmos. i Okeana, **27**, 3, 243–253 (1991)].

Stern, M. E. [1975] "Minimal properties of planetary eddies", J. Marine Res., **33**, 1, 1–13.

Syrovatsky, S. I. [1954] "Vortex sheet instability in a compressible medium", Zh. Eks. Teor. Fiz., **27**, 1, 121–123 (in Russian).

Tanaka, K., Ishii, S. [1982] "Scattering of a plane sound wave by a vortex pair", J. Phys. Soc. Jap., **51**, 6, 1992–1999.

Tatarsky, V. I. [1967] The Propagation of Waves in Turbulent Atmosphere, Nauka, Moscow (in Russian).

Thorpe, S. A. [1969] "Neutral eigensolutions of the stability equation for stratified shear flow", J. Fluid Mech., **36**, 4, 673–683.

Thorpe, S. A. [1984] "A laboratory study of stratified accelerating shear flow over a rough boundary", J. Fluid Mech., **318**, 185–196.

Timofeev, A. V. [1971] "Vibrations of inhomogeneous flows of plasmas and fluids", Sov. Phys. Usp., **13**, 632 [Engl. transl. from Usp. Fiz. Nauk, **102**, 2, 185 (1970)].

Timofeev, A. V. [1979] "On the nature of Rayleigh instability", Fizika Plazmy, **5**, 3, 705–706 (in Russian).

Timofeev, A. V. [1979] "On the theory of Alfven oscillations of inhomogeneous plasma", in: Problems of Plasma Theory, V. 9, Atomizdat, Moscow, pp. 205–232 (in Russian).

Tomita, H. [1986] "On nonlinear sea waves and the induced mean flow", J. Oceanogr. Soc. Jap., **42**, 153–160.

Troitskaya, Yu. I., Fabrikant, A. L. [1989] "Resonant amplification of internal gravity waves in a shear flow", Radiophys. and Quantum Electron., **32**, 10, 898–907 [Engl. transl. from Izv. VUZov, Radiofizika, **32**, 10, 1221–1231 (1989)].

Tsakadze, J. S., Tsakadze, S. J. [1975] "Superfluidity in pulsars", Usp. Fiz. Nauk., **115**, 3, 503–520 (in Russian).

Tsimring, L. Sh. [1982] "Destabilization of a stable shear flow under the action of internal waves", Okeanologiya, **22**, 4, 540–544 (in Russian).

Tsimring, L. Sh. [1983] "Stimulated scattering of surface wind waves", Izv. Atmos. Ocean. Phys., **19**, 68 [Engl. transl. from Izv. Akad. Nauk SSSR, Fiz. Atmos. i Okeana, **19**, 1, 68–73 (1983)].

Turner, J. S. [1973] Buoyancy Effects in Fluids, Cambridge University Press.

Uberoi, M. S. [1979] "Mechanism of decay of laminar and turbulent vortices", J. Fluid Mech., **90**, 2, 241–255.

Uberoi, M. S., Show, C. Y., Narain, J. P. [1972] Radio Sci., **15**, 10, 1718–1727.

Vainstein, L. A. [1966] Theory of Diffraction and Factorization Method, Sov. Radio, Moscow (in Russian).

Valenzuela, G. R. [1976] "The growth of gravity-capillary waves in a coupled shear flow", J. Fluid Mech., **76**, 2, 229–250.

Van Duin, C. A., Kelder, H. [1982] "Reflection properties of internal gravity waves incident upon a hyperbolic tangent shear layer", J. Fluid Mech., **120**, 505–521.

Van Dyke, M. [1975] Perturbation Methods in Fluid Mechanics, The Parabolic Press, Stanford, California.

Van Kampen, N. G. [1955] "On the theory of stationary waves in plasmas", Physica, **21**, 12, 949–963.

Vol'mir, A. C. [1976] Shells in Fluid Flows: Problems of Aeroelasticity, Nauka, Moscow, 416 pp. (in Russian).

Vol'mir, A. C. [1979] Shells in Fluid Flows: Problems of Hydroelasticity, Nauka, Moscow, 320 pp. (in Russian).

Volosov, V. M. [1962] "Averaging in systems of ordinary differential equations", Uspekhi Mat. Nauk, **17**, 6, 1541 (in Russian).

Voronovich, A. G., Rybak, S. A. [1978] "Explosive instability of stratified flows", Dokl. Akad. Nauk SSSR, **239**, 6, 1457–1460 (in Russian).

Vrobel', A. O., Kontorovich, V. M. [1982] "Wind instability of clouds of radiogalaxies", Sov. Astron. Lett., **8**, 179 [Engl. transl. from Pis'ma v Astronom. Zhurn., **8**, 6, 330–336 (1982)].

Wasow, W. [1965] Asymptotic Expansions for Ordinary Differential Equations, Interscience Publ., New York.

Whitham, G. B. [1974] Linear and Nonlinear Waves, Wiley Interscience Publ., John Wiley and Sons, N. Y.

Willoughby, H. E. [1977] "Inertia-buoyancy waves in hurricanes", J. Atmos. Sci., **34**, 7, 1028–1039.

Yamada, M., Goto, K. [1983] "Over-reflection of Rossby waves", J. Atmos. Sci., **40**, 2319–2321.

Yamada, M., Okamura, M. [1984] "Over-reflection and over-transmission of Rossby waves", J. Atmos. Sci., **41**, 16, 2531–2535.

Yuen, H. C., Lake, B. M. [1982] Nonlinear Dynamics of Deep-Water Gravity Waves, Academic Press, New York – London – Paris – San Diego – São Paulo – Sydney – Tokyo – Toronto.

Zakharov, S. B. [1976] "Stability of plane vortex layers in an ideal fluid", Uchenye Zapiski TsAGI, **7**, 3, 26–31 (in Russian).

Zakharov, V. E., Manakov, S. V., Novikov, S. P., Pitayevsky, L. P. [1980] Theory of Solitons. The Method of the Inverse Scattering Problem, Nauka, Moscow (in Russian).

Zakharov, V. E., Shabat, A. B. [1972] "Exact theory of two-dimensional self-focusing and one-dimensional self-modulation of waves in nonlinear media", Sov. Phys. JETP, **34**, 62–69 [Engl. transl. from Zh. Eksp. Teor. Fiz., **61**, 1, 118–134 (1971)].

Zarembo, L. K., Krasil'nikov, V. A. [1966] Introduction to Nonlinear Acoustics, Nauka, Moscow (in Russian).

Zel'dovich, Ya. B. [1972] "The amplification of cylindrical electromagnetic waves on reflection at a rotating body", Zh. Eksp. Teor. Fiz., **62**, 6, 2076–2081 (in Russian).

Zheleznyakov, V. V. [1977] Electromagnetic Waves in Space Plasma, Nauka, Moscow, 432 pp. (in Russian).

Zheleznyakov, V. V., Kocharovsky, V. V., Kocharovsky, Vl. V. [1984] "The effect of superradiation and dissipative instability in inverted two-level medium", Sov. Phys. JETP, **60**, 897 [Engl. transl. from Zh. Eksp. Teor. Fiz., **87**, 5, 1565–1581 (1984)].

Zhu, X., Holton, J. R. [1987] "Mean fields by local gravity-wave forcing in the middle atmosphere", J. Atmos. Sci., **44**, 3, 620–630.

The authors:

Anatoly L. Fabrikant: Ph. D. in Nonlinear Dynamics (Gorky State University, 1980) and Doctor of Sci. in Geophysical Fluid Dynamics (Institute of Applied Physics of the Russian Academy of Sciences, Nizhny Novgorod, 1991). Senior Scientist at the Department of Astrophysics and Space Physics in the Institute of Applied Physics, Russian Academy of Sciences (1977–1992). Currently works for Therma Wave, Inc., California, USA. Scientific interests are in wave theory, nonlinear dynamics, dynamics of the atmosphere and ocean, satellite remote sensing, acoustical and radio-acoustical sounding methods.

Yuri A. Stepanyants: Ph. D. in Geophysical Fluid Dynamics (Marine Hydrophysical Institute of the Ukrainian Academy of Sciences, Sebastopol, 1983) and Doctor of Sci. in Geophysical Fluid Dynamics (Institute of Applied Physics of the Russian Academy of Sciences, Nizhny Novgorod, 1992). Professor of Applied Mathematics at the Nizhny Novgorod State Technical University and Leading Researcher of the Department of Nonlinear Dynamics and Optics at the Institute of Applied Physics, Russian Academy of Sciences. Scientific interests are in geophysical fluid dynamics, oceanology, wave theory, nonlinear dynamics, applied mathematics.

